D0934328

# SUMMABILITY THEORY
# AND ITS APPLICATIONS

# THE NEW UNIVERSITY MATHEMATICS SERIES

*Editors:* Professor E. T. Davies, *Department of Mathematics, University of Waterloo,* and Professor J. V. Armitage, *Shell Centre for Mathematical Education, University of Nottingham*

This series is intended for readers whose main interest is in mathematics, or who need the methods of mathematics in the study of science and technology. Some of the books will provide a sound treatment of topics essential in any mathematical training, while other, more advanced, volumes will be suitable as preliminary reading for research in the field covered. New titles will be added from time to time.

BRICKELL and CLARK: *Differentiable Manifolds: An Introduction*
BROWN and PAGE: *Elements of Functional Analysis*
BURGESS: *Analytical Topology*
COOPER: *Functions of a Real Variable*
CURLE and DAVIES: *Modern Fluid Dynamics* (Vols 1 and 2)
EASTHAM: *Theory of Ordinary Differential Equations*
KUPERMAN: *Approximate Linear Algebraic Equations*
MAUNDER: *Algebraic Topology*
PORTEOUS: *Topological Geometry*
POWELL and SHAH: *Summability Theory and its Applications*
ROACH: *Green's Functions: Introductory Theory with Applications*
RUND: *The Hamilton–Jacobi Theory in the Calculus of Variations*
SMITH: *Introduction to the Theory of Partial Differential Equations*
SMITH: *Laplace Transform Theory*
SPAIN: *Ordinary Differential Equations*
SPAIN: *Vector Analysis*
SPAIN and SMITH: *Functions of Mathematical Physics*
ZAMANSKY: *Linear Algebra and Analysis*

# SUMMABILITY THEORY
# AND ITS APPLICATIONS

## R. E. POWELL
*Kent State University*

## S. M. SHAH
*University of Kentucky*

VAN NOSTRAND REINHOLD COMPANY
LONDON

NEW YORK   CINCINNATI   TORONTO   MELBOURNE

VAN NOSTRAND REINHOLD COMPANY LIMITED
25–28 Buckingham Gate, London S.W.1

INTERNATIONAL OFFICES
New York   Cincinnati   Toronto   Melbourne

Library of Congress Catalog Card No. 78-188530
ISBN 0 442 06625 2

*First published in 1972*

Printed in Great Britain by
Butler and Tanner Ltd
Frome and London

# Preface

The theory of summability has many uses throughout analysis and applied mathematics. The student of mathematics should have a basic understanding of and working ability in the area of summability. The engineer or physicist who deals with Fourier series, Fourier transforms, or analytic continuation will find the concepts of summability theory invaluable to his research.

The material in this text provides a classical introduction to summability theory and explores two particular applications; namely, applications to Fourier series and Fourier transforms and to analytic continuation. Since the exposition is classical in nature there is no prerequisite of functional analysis for the successful undertaking of the book. A student who has completed a rigorous year of intermediate analysis could read the first four chapters, while a student with a background in real and complex analysis should find all the chapters readily within his grasp.

The first chapter is primarily motivational and contains some basic notation and definitions. The Cesàro method of order $1((C,1)$ method) and the continuous Abel transform are examined here.

In the second chapter theorems concerning the general theory of matrix transformations are examined. The Silverman–Toeplitz theorem is proved and an alternate proof to this theorem based upon functional analytic techniques is given in the appendix.

Some standard summability transforms are investigated in the third chapter. Included are the Nörlund, Hölder, Cesàro (of order $\alpha$), Euler, Borel, Taylor and Hausdorff means.

The fourth chapter investigates the topic of Tauberian theorems. Tauber's original theorem is examined and various improvements and alternative approaches are studied.

Applications of summability theory are dealt with in the last two chapters. In the fifth chapter there is a self-contained introduction to Fourier series. Besides the standard tests of convergence, this chapter contains applications of Cesàro and Abel methods of summability to Fourier series. A brief account of the Riemann method of summation is also given. The Weierstrass theorems on approximation of a continuous function by (i) a trigonometric polynomial and (ii) an algebraic polynomial are proved. Also, estimates for errors in approximating a function by a trigonometric polynomial are given.

In the last chapter we investigate the minimal domain into which a summability transform continues analytically the sequence of partial sums of a power series expanded at the origin having positive radius of convergence. The Borel method is first examined and then the more general setting is presented, concluding with the Okada theorem.

In preparing this manuscript we have borrowed freely from various sources the ideas which we consider the simplest and the best. Throughout the book we provide complete and sufficiently detailed proofs to theorems and, on occasions, we have given alternate proofs of some standard theorems (such as the Riemann-Lebesgue theorem in Fourier series and the Silverman-Toeplitz theorem in the general theory) to expose the reader to different techniques of proof.

In order to keep the elementary level of presentation and so as not to increase the size of the book we have omitted certain topics such as the Riesz summability method, applications to Dirichlet series, Wierner's Tauberian theorems, and a proof of the prime number theorem. We refer the interested reader to the books of G. H. Hardy and M. Riesz, K. Chandrasekharam and S. Minakshisundaram, N. Wierner, G. H. Hardy, and H. R. Pitt found in the Bibliography.

Many people were involved in the writing and publishing of this book. Among these were students in seminars at the University of Kansas, Kent State University, and the University of Kentucky. Our heartfelt thanks go to them for their interest in the material and their criticisms of the presentation. Among our colleagues we wish to single out Professor John A. Fridy for his helpful suggestions. To Mrs. Julie Froble we extend our deepest appreciation for her patience and expert skill in preparing the manuscript. Finally, it is a pleasure to thank the publishers, Van Nostrand Reinhold Company Limited, and, in particular, Mr. D. J. Carpenter, Senior Editor, for the cooperation extended us and the care in the production of this book.

# CONTENTS

# CHAPTER 1

## Introduction

The study of convergence of series is an ancient art. The work done prior to the time of Léonard Euler (1707-83), was concerned essentially with orthodox examinations of convergence, and those series which did converge were of little or no interest. In fact such series were considered to be illegitimate (especially with respect to any emperical thinking of the real world) and, hence, one would be wise to leave them alone.

Euler developed the convention that a divergent series $\sum_{n=0}^{\infty} a_n = s$ provided $\sum_{n=0}^{\infty} a_n z^n$ converged to $f(z)$ for small values of $z$ and $f(1) = s$.

In this way $\dfrac{1}{(1+z)} = \sum_{n=0}^{\infty} (-1)^n z^n$ converges for $|z| < 1$ and, thus,

$$\sum_{n=0}^{\infty} (-1)^n = \left. \frac{1}{1+z} \right|_{z=1} = \frac{1}{2}.$$

However,
$$\frac{1-z^2}{1-z^3} = 1 - z^2 + z^3 - z^5 + z^6 - \ldots$$

and
$$\frac{1-z^2}{1-z^3} = \frac{1+z}{1+z+z^2} \quad \text{for } z \neq 1.$$

Thus
$$\frac{1+z}{1+z+z^2} = 1 - z^2 + z^3 - z^5 + z^6 - \ldots$$

and
$$\sum_{n=0}^{\infty} (-1)^n = \left. \frac{1+z}{1+z+z^2} \right|_{z=1} = \frac{2}{3}.$$

In this manner it would appear that one could assign almost any value to $\sum_{n=0}^{\infty} (-1)^n$ under Euler's interpretation.

Carl Friedrich Gauss (1777-1855) was instrumental in introducing the concept of the use of infinite processes into mathematical analysis. The binomial theorem (i.e. the expansion of $(1+x)^n$ where $n$ is not necessarily an integer) was mastered by Gauss at an early age and this got him interested in the concept of convergence of a power series to a function. This, of course, removed some of the absurdities encountered previously by mathematicians. These absurdities (such as we have already seen) were discarded as mysteries prior to this time. Gauss made other contributions to the mathematical world during his lifetime, among these being a discussion of the hypergeometric series and the convergence of such a series.

Augustin-Louis Cauchy (1789-1857) was a pioneer in the introduction of rigor into mathematical analysis (along with Gauss and Abel). He formalized ideas concerning convergence and divergence of power series. Much of his work still forms a basis for a good deal of present-day mathematical activity.

Niels Henrik Abel (1802-29) was a third important contributor to the ideas concerning convergence and divergence during the early part of the nineteenth century. The famous Theorem of Abel (see Theorem 1.5) is used many times over today.

The interest in divergent series (due, primarily, to Cauchy, Gauss, and Abel) declined in the second half of the nineteenth century only to be rekindled at a later date. Among those to begin re-examination of divergent series was E. Cesàro who, in 1890, introduced the idea of $(C,1)$ convergence (the reference is given in Chapter 3 following Theorem 3.5). This allowed mathematicians to, among other things, discuss the convergence of the Cauchy product of two infinite series (see Theorem 1.12 and Corollary 1.13). Many other mathematicians have published extensive works in the study of divergent series since that time. Throughout the book we make reference to some of them. There are, of course, many mathematicians who have contributed much to the study of convergence and divergence of series and sequences whom we do not mention. In no way is this meant to detract from their efforts or their contributions.

Let us now examine two of the earlier efforts in considering series and sequences which diverge. They are those of Abel and Cesàro (we consider only the $(C,1)$ method of convergence in this chapter).

**Abel convergence**

The method of convergence described in Definition 1.1 is referred to as Abel convergence due to Abel's theorem on the continuity of power series. Poisson used this method in the summation of Fourier series (see Chapter 5) and the method is sometimes attributed to him.

**1.1 DEFINITION** *Let $\{a_n\}_0^\infty$ be a sequence of complex numbers. The sequence $\{a_n\}_0^\infty$ is Abel convergent (written (A) convergent) to L if*

$$\lim_{x \to 1-} (1-x) \sum_{k=0}^{\infty} a_k x^k = L \text{ exists}$$

*(where $\lim_{x \to 1-} f(x) = M$ if given $\varepsilon > 0$ there exists $\delta = \delta(\varepsilon) > 0$ such that if $1 - \delta < x < 1$ then $|f(x) - M| < \varepsilon$).*

The natural question 'How does the ordinary convergence of a sequence and the (A) convergence of a sequence compare?' arises. The next theorem answers this question.

**1.2 THEOREM** *If $\{a_n\}_0^\infty$ converges to L then $\{a_n\}_0^\infty$ (A) converges to L (but not conversely).*

*Proof.* Let $f(x) = (1-x) \sum_{k=0}^{\infty} a_k x^k$ for $-1 < x < 1$. Let $0 < \varepsilon < 1$ be given.

given. Choose a positive integer $N = N(\varepsilon)$ such that if $n \geqslant N$ then $|a_n - L| < \frac{1}{2}\varepsilon$. Let $M = \max\{|a_0 - L|, \ldots, |a_N - L|\}$. We have

$(1-x) \sum_{k=0}^{\infty} x^k = 1$ for $-1 < x < 1$. Let $\delta = \dfrac{\varepsilon}{(N+1)(M+1)}$. If $x \in$

$(1-\delta, 1)$ then

$$|f(x) - L| = |(1-x) \sum_{k=0}^{\infty} a_k x^k - L|$$

$$= |(1-x) \sum_{k=0}^{\infty} a_k x^k - (1-x) \sum_{k=0}^{\infty} L x^k|$$

$$= |(1-x) \sum_{k=0}^{\infty} (a_k - L) x^k|$$

$$\leqslant |(1-x) \sum_{k=0}^{N} (a_k - L) x^k| + |(1-x) \sum_{k=N+1}^{\infty} (a_k - L) x^k|$$

$$< (1-x)(N+1) M + \frac{1}{2}\varepsilon = \frac{\varepsilon}{2(N+1)(M+1)} (N+1) M + \frac{1}{2}\varepsilon$$

$$< \frac{1}{2}\varepsilon + \frac{1}{2}\varepsilon = \varepsilon.$$

Therefore, $\lim_{x \to 1-} f(x) = \lim_{x \to 1-} (1-x) \sum_{k=0}^{\infty} a_k x^k = L$, i.e. the sequence $\{a_n\}_0^\infty$ is (A) convergent to $L$.

To see that the converse does not hold let $a_n = 1 + (-1)^n$ for $n = 0, 1, \ldots$ . The sequence $\{a_n\}_0^\infty$ diverges, however,

$$\lim_{x \to 1-} (1-x) \sum_{k=0}^{\infty} a_k x^k = \lim_{x \to 1-} \sum_{k=0}^{\infty} 2(-1)^k x^k = 2 \lim_{x \to 1-} \frac{1}{1+x} = 1. \quad \triangle$$

Therefore, we have that convergence of $\{a_n\}_0^{\infty}$ to $L$ implies $(A)$ convergence of $\{a_n\}_0^{\infty}$ to $L$ but not conversely (in fact, $(A)$ convergence does not even imply convergence).

In order to prove a generalization of Theorem 1.2 we need the following definition and lemma.

1.3 DEFINITION. *Let* $0 < \alpha < \frac{1}{2}\pi$. *A Stolz domain of angle* $\alpha$, *written* $S(\alpha)$, *is the domain*

$$\{w: |\arg (1-w)| < \alpha\} \cap S_1(0) \text{ (where } S_r(w) = \{z: |z-w| < r\}).$$

1.4 LEMMA. *Let* $0 < \alpha < \frac{1}{2}\pi$. *If* $z \in S(\alpha) \cap S_R(1)$ *where* $R = \cos \alpha$ *then*

$$\frac{|1-z|}{1-|z|} \leqslant \frac{2}{\cos \alpha}.$$

*Proof.* Let $z = 1 - r(\cos \theta + i \sin \theta)$ where $|\theta| < \alpha < \frac{1}{2}\pi$ and $0 < r < R = \cos \alpha$. We have

$$\frac{|1-z|}{1-|z|} = \frac{r}{1 - (1 - 2r \cos \theta + r^2)^{1/2}} \leqslant \frac{2}{\cos \alpha}$$

if and only if $r \cos \alpha \leqslant 2[1 - (1 - 2r \cos \theta + r^2)^{1/2}]$ and this holds if and only if

$$[(1 - 2r \cos \theta + r^2)^{1/2}]^2 \leqslant (1 - \tfrac{1}{2} r \cos \alpha)^2,$$

i.e.                  $-2 \cos \theta + r \leqslant -\cos \alpha + \tfrac{1}{4} r \cos^2 \alpha.$

Since $|\theta| < \alpha < \frac{1}{2}\pi$ and $0 < r < \cos \alpha$ we have

$$-2 \cos \theta + r \leqslant -2 \cos \alpha + \cos \alpha = -\cos \alpha \leqslant -\cos \alpha + \tfrac{1}{4} r \cos^2 \alpha.$$

Therefore,

$$\frac{|1-z|}{1-|z|} \leqslant \frac{2}{\cos \alpha} \quad \text{for } z \in S(\alpha) \cap S_{\cos \alpha}(1). \quad \triangle$$

1.5 THEOREM. (A generalization of Theorem 1.2.) *Let* $0 < \alpha < \frac{1}{2}\pi$. *If the sequence* $\{a_n\}_0^{\infty}$ *converges to* $L$ *then*

$$\lim_{z \to 1, \, z \in S(\alpha)} (1-z) \sum_{k=0}^{\infty} a_k z^k = L.$$

*Proof.* Let $\varepsilon > 0$ be given. For $z \in S(\alpha) \cap S_{\cos \alpha}(1)$ we have, by Lemma 1.4, that $(|1-z|)/(1-|z|) \leqslant 2/\cos \alpha$. Since $\{a_n\}_0^{\infty}$ converges to $L$ we can choose a positive integer $N = N(\varepsilon)$ such that $|a_n - L| < (\varepsilon \cos \alpha)/4$ whenever $n \geqslant N$. Define $M = \max \{|a_0 - L|, \ldots, |a_{N-1} - L|\}$.

Choose $\delta = \min \{\cos \alpha, \ \varepsilon/[2(M+1) \ (N+1)]\}$.

If $z \in S(\alpha) \cap S_\delta(1)$ then

$$\left|(1-z) \sum_{k=0}^\infty a_k z^k - L \right| \ = \ \left|(1-z) \sum_{k=0}^\infty a_k z^k - (1-z) \sum_{k=0}^\infty L z^k \right|.$$

Therefore, $\qquad\qquad \left|(1-z) \sum_{k=0}^\infty a_k z^k - L \right|$

$$= \ \left| 1-z \right| \ \left| \sum_{k=0}^\infty (a_k - L) z^k \right|$$

$$\leqslant \ \left| 1-z \right| \ [\, \sum_{k=0}^{N-1} \left| a_k - L \right| \ |z|^k + \sum_{k=N}^\infty \left| a_k - L \right| \ |z|^k \,]$$

$$\leqslant \ \left| 1-z \right| NM + \left| 1-z \right| \frac{\varepsilon \cos \alpha}{4} \sum_{k=0}^\infty |z|^k$$

$$< \ \frac{\varepsilon}{2(M+1) \ (N+1)} NM + \frac{\varepsilon \cos \alpha}{4} \frac{|1-z|}{1-|z|}$$

$$< \ \tfrac{1}{2} \varepsilon + \frac{\varepsilon \cos \alpha}{4} \frac{2}{\cos \alpha} = \tfrac{1}{2} \varepsilon + \tfrac{1}{2} \varepsilon \ = \ \varepsilon.$$

Therefore, $\quad \lim_{z \to 1, \ z \in S(\alpha)} (1-z) \sum_{k=0}^\infty a_k z^k \ = \ L. \ \triangle$

Of course, the converse to Theorem 1.5 does not hold either (use the same counter example as we used in the proof of Theorem 1.2).

**1.6 DEFINITION.** *A series of complex numbers,* $\sum_{k=0}^\infty a_k$, *is* (A) *convergent to* L *if the sequence of partial sums,* $\{s_n\}_0^\infty$ *(where* $s_n = \sum_{k=0}^n a_k$), *is* (A) *convergent to* L.

**1.7. THEOREM.** *If the series* $\sum_{k=0}^\infty a_k = M$ *then* $\lim_{x \to 1-} \sum_{k=0}^\infty a_k x^k = M$.

*Proof.* Let $s_n = \sum_{k=0}^n a_k$. For $|x| < 1$ the series $\sum_{k=0}^\infty a_k x^k$ is absolutely convergent. Since $\{s_n\}_0^\infty$ is convergent to $M$ we have, by Theorem 1.2, that $\{s_n\}_0^\infty$ is (A) convergent to $M$, i.e.

$$M \ = \ \lim_{x \to 1-} (1-x) \sum_{k=0}^\infty s_k x^k = \lim_{x \to 1-} [\, \sum_{k=0}^\infty s_k x^k - \sum_{k=0}^\infty s_k x^{k+1} \,]$$

$$= \ \lim_{x \to 1-} [\, s_0 + \sum_{k=1}^\infty s_k x^k - \sum_{k=1}^\infty s_{k-1} x^k \,]$$

$$= \lim_{x \to 1-} [s_0 + \sum_{k=1}^{\infty} (s_k - s_{k-1}) x^k] = \lim_{x \to 1-} [a_0 + \sum_{k=1}^{\infty} a_k x^k]$$

$$= \lim_{x \to 1-} \sum_{k=0}^{\infty} a_k x^k. \quad \triangle$$

*Exercise.* Prove the statement: If $\sum_{k=0}^{\infty} a_k$ is a series of complex numbers then $\sum_{k=0}^{\infty} a_k$ is (A) convergent if and only if $\lim_{x \to 1-} \sum_{k=0}^{\infty} a_k x^k$ exists.

**1.8 EXAMPLE** We have that $\sum_{k=0}^{\infty} z^k = 1/(1-z)$ for $|z| < 1$. Let $s_n(z) = \sum_{k=0}^{n} z^k$ and consider the (A) convergence of $\{s_n(z)\}_0^{\infty}$. For $z \neq 1$ we have $s_n(z) = (1-z^{n+1})/(1-z)$ and $\lim_{x \to 1-} (1-z) \sum_{k=0}^{\infty} s_k(z) x^k$

$$= \lim_{x \to 1-} (1-x) \sum_{k=0}^{\infty} \frac{1-z^{k+1}}{1-z} x^k$$

$$= \lim_{x \to 1-} \left[ \frac{1-x}{1-z} \sum_{k=0}^{\infty} x^k - \frac{z(1-x)}{1-z} \sum_{k=0}^{\infty} (zx)^k \right]$$

$$= \frac{1}{1-z} - \lim_{x \to 1-} \frac{z(1-x)}{(1-z)(1-zx)}$$

provided $|zx| < 1$. Therefore, if $z \neq 1$ then $\{s_n(z)\}_0^{\infty}$ is (A) convergent if and only if $\lim_{x \to 1-} (1-x)/(1-zx)$ exists and $|z| \leqslant 1$. This limit exists and is 0 for $z \neq 1$. Therefore, if $|z| \leqslant 1$ and $z \neq 1$ then $\{s_n(z)\}_0^{\infty}$ is (A) convergent (to $1/(1-z)$ in fact) and conversely.

*Exercise.* Prove: If $\lim_{n \to \infty} a_n = +\infty$ and $\sum_{k=0}^{\infty} a_k x^k$ converges for $|x| < 1$ then

$$\lim_{x \to 1-} (1-x) \sum_{k=0}^{\infty} a_k x^k = +\infty.$$

**Cesàro Convergence**
**1.9 DEFINITION.** *Let* $\{a_n\}_0^{\infty}$ *be a sequence of complex numbers. The sequence* $\{a_n\}_0^{\infty}$ *is Cesàro convergent (written (C,1) convergent) to* $L$ *if* $\lim_{n \to \infty} \frac{1}{(n+1)} \sum_{k=0}^{\infty} a_k = L$ *exists.*

From Definition 1.9 we see that the idea of Cesàro convergence is to take the arithmetic mean of the terms comprising the sequence and study the convergence of these means.

We can compare (C,1) convergence not only to convergence, but also, to $(A)$ convergence. We will see that convergence implies $(C,1)$ convergence and $(C,1)$ convergence implies $(A)$ convergence, however, no two of the three are equivalent.

1.10 THEOREM. *If $\{a_n\}_0^\infty$ converges to $L$ then $\{a_n\}_0^\infty$ is $(C,1)$ convergent to $L$ (but not conversely).*

*Proof.* Let $\varepsilon > 0$ be given. There exists a positive integer $N_1$ such that $|a_n - L| < \frac{1}{2}\varepsilon$ for $n \geqslant N_1$. There exists a positive integer $N_2$ such that $\frac{1}{n} \mid [a_0 + \ldots + a_{N_1-1}] - N_1 L \mid < \frac{1}{2}\varepsilon$ for $n \geqslant N_2$. Let $N = \max\{N_1, N_2\}$. If $n \geqslant N$ then $\mid \frac{1}{n+1} \sum_{k=0}^{n} a_k - L \mid$

$$= \mid \frac{1}{n+1} [\sum_{k=0}^{n} a_k - (n+1) L] \mid$$

$$\leqslant \frac{1}{n+1} \mid \sum_{k=0}^{N_1-1} a_k - N_1 L \mid + \frac{1}{n+1} \mid \sum_{k=N_1}^{n} a_k - (n+1-N_1) L \mid$$

$$< \frac{1}{2}\varepsilon + \frac{1}{n+1} \mid \sum_{k=N_1}^{n} [a_k - L] \mid \leqslant \frac{1}{2}\varepsilon + \frac{1}{n+1} \sum_{k=N_1}^{n} \mid a_k - L \mid$$

$$< \frac{1}{2}\varepsilon + \frac{1}{n+1} (n+1-N_1) \frac{1}{2}\varepsilon \leqslant \frac{1}{2}\varepsilon + \frac{1}{2}\varepsilon = \varepsilon.$$

The fact that the converse does not hold follows from the example $a_n = 1 + (-1)^n$ for $n = 0, 1, \ldots$ . The sequence $\{a_n\}_0^\infty$ diverges however, $\{a_n\}_0^\infty$ is $(C,1)$ convergent to 1 as we now prove. Let $\varepsilon < 0$ be given. We have that

$$\sum_{k=0}^{n} a_k = \begin{cases} n+2 & \text{if } n \text{ is even} \\ \\ n+1 & \text{if } n \text{ is odd.} \end{cases}$$

Therefore, $\mid \frac{1}{n+1} \sum_{k=0}^{n} a_k - 1 \mid \leqslant \left(\frac{n+2}{n+1} - 1\right) = \frac{1}{n+1}$.

Choose a positive integer $N(\varepsilon) > (1-\varepsilon)/\varepsilon$. If $n \geqslant N(\varepsilon)$ then $1/(n+1) < \varepsilon$, i.e. if $n \geqslant N(\varepsilon)$ then $\mid \frac{1}{n+1} \sum_{k=0}^{n} a_k - 1 \mid < \varepsilon$. $\triangle$

1.11 DEFINITION. *A series, $\sum_{k=0}^{\infty} a_k$, of complex numbers is $(C,1)$ convergent to $L$ if the sequence of partial sums, $\{s_n\}_0^\infty$ (where $s_n = \sum_{k=0}^{n} a_k$) is $(C,1)$ convergent to $L$.*

The following theorem is very useful when we are studying Cauchy products of series. Recall that the Cauchy product of the series $\sum_{k=0}^{\infty} a_k$ and $\sum_{k=0}^{\infty} b_k$ is the series $\sum_{k=0}^{\infty} c_k$ where $c_k = \sum_{j=0}^{k} a_j b_{k-j}$ for $k = 0, 1, \ldots$ and that if $\sum_{k=0}^{\infty} a_k = A$, $\sum_{k=0}^{\infty} b_k = B$, and one of these series is absolutely convergent then $\sum_{k=0}^{\infty} c_k = AB$. Now we prove

**1.12 THEOREM.** *If* $\sum_{k=0}^{\infty} a_k = A$ *and* $\sum_{k=0}^{\infty} b_k = B$ *then the Cauchy product,* $\sum_{k=0}^{\infty} c_k$, *is* (C,1) *convergent to* $AB$.

*Proof.* Let $A_n = \sum_{k=0}^{n} a_k$, $B_n = \sum_{k=0}^{n} b_k$, $s_n = \sum_{k=0}^{n} c_k$, and $E_n = \dfrac{1}{n+1} \sum_{k=0}^{n} s_k$.

We have

$$s_n = \sum_{k=0}^{n} c_k = \sum_{k=0}^{n} \left[ \sum_{j=0}^{k} a_j b_{k-j} \right] = \sum_{j=0}^{n} \left[ \sum_{k=j}^{n} a_j b_{k-j} \right]$$

$$= \sum_{j=0}^{n} a_j \left[ \sum_{k=j}^{n} b_{k-j} \right] = \sum_{j=0}^{n} a_j B_{n-j}.$$

Also, 

$$E_n = \frac{1}{n+1} \sum_{k=0}^{n} s_k = \frac{1}{n+1} \sum_{k=0}^{n} \left[ \sum_{j=0}^{k} a_j B_{k-j} \right]$$

$$= \frac{1}{n+1} \sum_{k=0}^{n} \left[ \sum_{j=0}^{k} a_{k-j} B_j \right]$$

$$= \frac{1}{n+1} \sum_{j=0}^{n} B_j \left[ \sum_{k=j}^{n} a_{k-j} \right] = \frac{1}{n+1} \sum_{j=0}^{n} B_j A_{n-j}.$$

So, 

$$E_n = \frac{1}{n+1} \sum_{j=0}^{n} B_j (A_{n-j} - A) + \frac{1}{n+1} \sum_{j=0}^{n} B_j A$$

$$= \frac{1}{n+1} \sum_{j=0}^{n} B_j (A_{n-j} - A) + A \left[ \frac{1}{n+1} \sum_{j=0}^{n} B_j \right].$$

Since $\{B_k\}_0^{\infty}$ converges to $B$ we have, by Theorem 1.10, that $\lim_{n \to \infty} \dfrac{1}{n+1} \sum_{j=0}^{n} B_j = B$. Therefore, $\lim_{n \to \infty} E_n = AB$ if and only if $\lim_{n \to \infty} \dfrac{1}{n+1} \sum_{j=0}^{n} B_j (A_{n-j} - A) = 0$.

Let $\varepsilon > 0$ be given. Since $\{B_n\}_0^\infty$ and $\{A_n - A\}_0^\infty$ each converge they are bounded, say $|B_n| \leqslant M_1$ for all $n = 0, 1, \ldots$ and $|A_n - A| \leqslant M_2$ for all $n = 0, 1, \ldots$ for some $M_1 > 0$ and $M_2 > 0$. Also, $\{A_n - A\}_0^\infty$ converges to 0 so there exists a positive integer $N_1 = N_1(\varepsilon)$ such that if $n \geqslant N_1$ then $|A_n - A| < \varepsilon/2M_1$. There exists a positive integer $N_2$ such that $(N_1+1)/N_2 < \varepsilon/2M_1 M_2$. Let $N = \max \{N_1, N_2\}$. If $n \geqslant N$ then $\left| \dfrac{1}{n+1} \sum\limits_{j=0}^{n} B_j (A_{n-j} - A) \right|$

$$= \frac{1}{n+1} \left| \sum_{j=0}^{n} B_{n-j} (A_j - A) \right|$$

$$\leqslant \frac{1}{n+1} \sum_{j=0}^{N_1} |B_{n-j}| \, |A_j - A| + \frac{1}{n+1} \sum_{j=N_1+1}^{n} |B_{n-j}| \, |A_j - A|$$

$$< \frac{1}{n+1} \sum_{j=0}^{N_1} M_1 M_2 + \frac{1}{n+1} \sum_{j=N_1+1}^{n} M_1 \left( \frac{\varepsilon}{2M_1} \right)$$

$$= \frac{N_1+1}{n+1} M_1 M_2 + \frac{n-N_1}{n+1} \frac{1}{2} \varepsilon < \frac{\varepsilon}{2M_1 M_2} (M_1 M_2) + \frac{1}{2}\varepsilon = \frac{1}{2}\varepsilon + \frac{1}{2}\varepsilon = \varepsilon. \quad \triangle$$

**1.13 COROLLARY.** *If $\sum\limits_{k=0}^{\infty} a_k = A$, $\sum\limits_{k=0}^{\infty} b_k = B$, and the Cauchy product of the two series $\sum\limits_{k=0}^{\infty} c_k = C$ then $C = AB$.*

*Proof.* The series $\sum\limits_{k=0}^{\infty} c_k$ is $(C,1)$ convergent to $AB$ by Theorem 1.12 and, by Theorem 1.10, $\sum\limits_{k=0}^{\infty} c_k$ is $(C,1)$ convergent to $C$. Since the limits of sequences are unique and the sequence $\{E_n\}_0^\infty$ converges to both $AB$ and $C$ (where $E_n$ is defined in the proof of Theorem 1.12) we have that $AB = C$. $\triangle$

Corollary 1.13 says, essentially, if the Cauchy product of two convergent series converges then it converges to the 'correct' value.

We conclude this section by comparing the $(A)$ convergence of a sequence with the $(C,1)$ convergence of the sequence.

**1.14 THEOREM.** *If the sequence $\{a_n\}_0^\infty$ is $(C,1)$ convergent to $L$ then $\{a_n\}_0^\infty$ is $(A)$ convergent to $L$ (but not conversely).*

*Proof.* Write $s_n = \dfrac{1}{n+1} \sum\limits_{k=0}^{n} a_k$ and let $f(x) = (1-x) \sum\limits_{k=0}^{\infty} a_k x^k$.

Now, $(n+1)s_n = \sum_{k=0}^{n} a_k$ and $ns_{n-1} = \sum_{k=0}^{n-1} a_k$.                Therefore,

$a_k = (k+1)s_k - ks_{k-1}$ and we have

$$f(x) = (1-x) \sum_{k=0}^{\infty} a_k x^k = (1-x) [a_0 + \sum_{k=1}^{\infty} [(k+1)s_k - ks_{k-1}] x^k].$$

The series $\sum_{k=0}^{\infty} (k+1)x^k$ has radius of convergence 1 (since $\sum_{k=0}^{\infty} x^{k+1}$

has radius of convergence 1) and, hence, the series $\sum_{k=0}^{\infty} (k+1)s_k x^k$

and $\sum_{k=1}^{\infty} ks_{k-1}x^k$ each have radius of convergence at least 1 (since

$\{s_k\}_0^{\infty}$ converges to a finite limit $L$). Therefore, $\sum_{k=1}^{\infty} [(k+1)s_k - ks_{k-1}]x^k$

has radius of convergence at least 1, and, hence, $(1-x) \sum_{k=0}^{\infty} a_k x^k$

converges for $|x| < 1$.   We want to show $\lim_{x \to 1-} (1-x) \sum_{k=0}^{\infty} a_k x^k$

$= \lim_{x \to 1-} f(x) = L$.

Now, $\dfrac{1}{(1-x)^2} f(x)$

$$= \frac{1}{1-x} \sum_{k=0}^{\infty} a_k x^k = [\sum_{k=0}^{\infty} x^k] [\sum_{k=0}^{\infty} a_k x^k]$$

$$= \sum_{k=0}^{\infty} (a_0 + \ldots + a_k) x^k = \sum_{k=0}^{\infty} (k+1) s_k x^k.$$

Also,          $\dfrac{1}{(1-x)^2} = \dfrac{d}{dx} \dfrac{1}{1-x} = \sum_{k=0}^{\infty} (k+1) x^k$

and, hence, $\dfrac{L}{(1-x)^2} = \sum_{k=0}^{\infty} (k+1) L x^k.$

Therefore, $|f(x) - L| = |(1-x)^2 \sum_{k=0}^{\infty} (k+1) (s_k - L) x^k|$

$\leqslant (1-x)^2 \sum_{k=0}^{\infty} (k+1) |s_k - L| x^k$  for $0 < x < 1$.

We have that $\{s_n\}_0^{\infty}$ converges to $L$ so, given $\varepsilon > 0$, there exists a positive integer $N = N(\varepsilon)$ such that if $n \geqslant N$ then $|s_n - L| < \frac{1}{2}\varepsilon$. Since $\{s_n\}_0^{\infty}$ is convergent, we have that it is bounded.   Suppose that $|s_n| \leqslant M$ for all $n = 0, 1, \ldots$ and for some $M > 0$.   Choose $\delta(\varepsilon) = \min\{\frac{1}{2}, [\varepsilon/4 (M+1) (N+1)^2]^{1/2}\}$.   If $1 - \delta(\varepsilon) < x < 1$ then

$$|f(x) - L| \leqslant (1-x)^2 \sum_{k=N}^{\infty} (k+1)|s^k - L|x^k + (1-x)^2 \sum_{k=0}^{N-1} (k+1)|s_k - L|x^k$$

$$< (1-x)^2 \tfrac{1}{2}\varepsilon \sum_{k=0}^{\infty} (k+1)x^k + \frac{\varepsilon}{4(M+1)(N+1)^2} N(2M) \sum_{k=0}^{N-1} x^k$$

$$< \frac{(1-x)^2}{(1-x)^2} \tfrac{1}{2}\varepsilon + \frac{2MN^2\varepsilon}{4(M+1)(N+1)^2} < \tfrac{1}{2}\varepsilon + \tfrac{1}{2}\varepsilon = \varepsilon.$$

Therefore, $\lim_{x \to 1-} f(x) = \lim_{x \to 1-} \sum_{k=0}^{\infty} a_k x^k = L$, i.e. the sequence $\{a_n\}_0^{\infty}$ is $(A)$ convergent to $L$.

To see that the converse of the theorem does not hold define

$$a_n = \begin{cases} k+1 & \text{if } n = 2k, \ k = 0, 1, \dots \\ -(k+1) & \text{if } n = 2k+1, \ k = 0, 1, \dots \end{cases}.$$

First we prove that $\{a_n\}_0^{\infty}$ is not $(C,1)$ convergent. Let $s_n = \frac{1}{n+1} \sum_{k=0}^{n} a_k$.

Then
$$s_n = \begin{cases} 0 & \text{if } n = 2k+1, \ k = 0, 1, \dots \\ \dfrac{k+1}{2k+1} & \text{if } n = 2k, \ k = 0, 1, \dots \end{cases}.$$

Therefore, $\{s_{2k+1}\}_0^{\infty}$ converges to $0$ and $\{s_{2k}\}_0^{\infty}$ converges to $\frac{1}{2}$. However, we have that $\{s_n\}_0^{\infty}$ does not converge.

Finally, we show that $\{a_n\}_0^{\infty}$ is $(A)$ convergent to $\frac{1}{4}$. Consider, for $|x| < 1$,

$$(1-x) \sum_{k=0}^{\infty} a_k x^k = \sum_{k=0}^{\infty} a_k x^k - \sum_{k=0}^{\infty} a_k x^{k+1}$$

$$= a_0 + \sum_{k=1}^{\infty} (a_k - a_{k-1}) x^k = 1 + \sum_{k=1}^{\infty} (-1)^k (k+1) x^k$$

$$= \sum_{k=0}^{\infty} (k+1)(-x)^k = \sum_{k=1}^{\infty} k(-x)^{k-1} = -\sum_{k=1}^{\infty} \frac{d}{dx}(-x)^k$$

$$= -\frac{d}{dx} \sum_{k=0}^{\infty} (-x)^k = -\frac{d}{dx} \frac{1}{1+x} = \frac{1}{(1+x)^2}$$

Therefore, $\lim_{x \to 1-} (1-x) \sum_{k=0}^{\infty} a_k x^k = \lim_{x \to 1-} \frac{1}{(1+x)^2} = \frac{1}{4}$. So, $\{a_n\}_0^{\infty}$ is not $(C,1)$ convergent but it is $(A)$ convergent. $\triangle$

## The Symbols $O$, $o$, $\sim$.

In this section we want to compare the growth of two sequences $\{a_n\}_0^\infty$ and $\{b_n\}_0^\infty$ as $n$ approaches $\infty$ or, more generally, of two functions $f(x)$ and $g(x)$ as $x$ tends to a limit $x_0$. To do this we introduce the symbols $O$, $o$, and $\sim$.

**1.15 DEFINITION.** *Let $f(x)$ and $g(x)$ be given functions. Let $x_0$ be a fixed point and suppose that $g(x)$ is positive and continuous in an open interval about $x_0$.*

    (i) *If there is a constant $K$ such that $|f(x)| < Kg(x)$ in an open interval about $x_0$ then $f(x) = O(g(x))$, $(x \to x_0)$.*

    (ii) *If $\lim_{x \to x_0} \dfrac{f(x)}{g(x)} = 0$ then $f(x) = o(g(x))$, $(x \to x_0)$.*

    (iii) *If $\lim_{x \to x_0} \dfrac{f(x)}{g(x)} = 1$ then $f(x) \sim g(x)$, $(x \to x_0)$.*

*In (iii) $g(x)$ does not have to be positive about $x_0$.*

In this definition $x_0$ may be finite or infinite (in the case where $x_0$ is infinite we replace the condition that $g(x)$ is positive and continuous in an open interval about $x_0$ with the condition that $g(x)$ is positive and continuous for $x$ large enough, i.e. there exists $\xi$ such that $g(x)$ is positive and continuous for $x > \xi$).

**1.16 EXAMPLE.** (i) $5x^2 + 2x + 3 = O(x^2)$, $(x \to \infty)$ since, for $x > 3$, $|5x^2 + 2x + 3| = 5x^2 + 2x + 3 < 6x^2$.

    (ii) $\sin x = o(x)$, $(x \to \infty)$ since $|\sin x| \leqslant 1$, $\lim_{x \to \infty} \dfrac{\sin x}{x} = 0$.

    (iii) $\sin x \sim x$, $(x \to 0)$ since $\lim_{x \to 0} \dfrac{\sin x}{x} = 1$.

    (iv) $1 - \cos x = o(x)$, $(x \to 0)$ since $0 \leqslant 1 - \cos x$
$$= 2 \sin^2 \frac{x}{2} \sim 2\left(\frac{x}{2}\right)^2 = \frac{x^2}{2} = o(x).$$

    (v) $x^n = o(e^x)$, $(x \to \infty)$ since $e^x = \sum_{k=0}^\infty \dfrac{x^k}{k!} > \dfrac{x^{n+1}}{(n+1)!}$ for $x > 0$

and, hence, $0 \leqslant \dfrac{x^n}{e^x} < \dfrac{(n+1)!}{x}$ which converges to 0 as $x \to \infty$.

If $f(x)$ is bounded as $x \to x_0$ we take $g(x) = 1$ in Definition 1.15 and write $f(x) = O(1)$, $(x \to x_0)$. If $\lim_{x \to x_0} f(x) = 0$ we write $f(x) = o(1)$, $(x \to x_0)$. Thus, if $b_1 \neq 0$ then

$$R(x) = \frac{a_1 x^m + a_2 x^{m-1} + \ldots + a_{m+1}}{b_1 x^m + b_2 x^{m-1} + \ldots + b_{m+1}} = O(1), \quad (x \to \infty).$$

If $a_1 = 0$ then $R(x) = o(1)$, $(x \to \infty)$. Finally, if $a_1 b_1 \neq 0$ then

$R(x) \sim a_1/b_1, \ (x \to \infty)$.

1.17 EXAMPLE. If $a_n = o(1), \ b_n = o(1)$ (both as $n \to \infty$), and if the sequence $\{b_n\}_0^\infty$ is decreasing then

$$\lim_{n \to \infty} a_n/b_n = \lim_{n \to \infty} (a_n - a_{n+1})/(b_n - b_{n+1})$$

provided the second quotient has a limit (either finite or infinite).

Suppose, first, this limit is finite (say the limit is $L$). Let $\varepsilon > 0$ be given. There exists a positive integer $N = N(\varepsilon)$ such that if $n \geqslant N$ then $L - \varepsilon < (a_n - a_{n+1})/(b_n - b_{n+1}) < L + \varepsilon$. So,

$$(L - \varepsilon)(b_n - b_{n+1}) < a_n - a_{n+1} < (L + \varepsilon)(b_n - b_{n+1}) \quad \text{for } n \geqslant N.$$

Now, let $n = k, k+1, \ldots, k+p$ in this inequality and sum these resulting inequalities. This yields

$$(L - \varepsilon)(b_k - b_{k+p}) < a_k - a_{k+p} < (L + \varepsilon)(b_k - b_{k+p}) \quad \text{for } k \geqslant N.$$

Now, let $p \to \infty$ in this inequality. This gives us

$$(L - \varepsilon) b_k \leqslant a_k \leqslant (L + \varepsilon) b_k \quad \text{for } k \geqslant N, \ \text{i.e. } \lim_{n \to \infty} \frac{a_n}{b_n} = L.$$

In the case when the limit is $\infty$ we have, given $H > 0$, that $(a_n - a_{n+1})/(b_n - b_{n+1}) > H$ for $n$ sufficiently large (say $n \geqslant N = N(H)$), i.e. $a_n - a_{n+1} > H(b_n - b_{n+1})$ for $n \geqslant N$. Now, as before, we obtain, for $k \geqslant N$, $a_k - a_{k+p} > H(b_k - b_{k+p})$ and the result follows by letting $p \to \infty$.

## A Necessary Condition for (C,1) Convergence

In the following material we will omit '$(n \to \infty)$' after the notation $o(n)$ or $O(n)$ if there can be no confusion as to what is meant.

Let $\sum_{k=0}^{\infty} a_k$ be a series of complex numbers and let $s_n = \sum_{k=0}^{n} a_k$. If $\sum_{k=0}^{\infty} a_k$ converges then we know that $a_n = o(1)$ and $s_n = O(1)$ (since a convergent sequence is bounded). We now see that $(C,1)$ convergence of the series gives us $s_n = o(n)$ and $a_n = o(n)$.

1.18 THEOREM. If $\sum_{k=0}^{\infty} a_k$ is $(C,1)$ convergent then $s_n = o(n)$ and $a_n = o(n)$.

Proof. Write $t_n = \dfrac{1}{n+1} \sum_{k=0}^{n} s_k$. Therefore, $s_n = (n+1)t_n - n t_{n-1}$. By hypothesis, $\{t_n\}_0^\infty$ converges (say to $L$). Therefore,

$\frac{s_n}{n} = (1 + \frac{1}{n}) t_n - t_{n-1} = (t_n - t_{n-1}) + \frac{1}{n} t_n.$  Therefore, we obtain,

$\lim_{n \to \infty} \frac{s_n}{n} = \lim_{n \to \infty} \left[ (t_n - t_{n-1}) + \frac{1}{n} t_n \right] = L - L + \lim_{n \to \infty} \frac{t_n}{n} = 0,$

i.e. $s_n = o(n).$

Furthermore, $a_n = s_n - s_{n-1} = o(n) + o(n) = o(n).$     $\triangle$

**1.19  EXAMPLE.**  The series $1 - 2 + 3 - 4 + \ldots + (-1)^{n-1} n + \ldots$ is not $(C,1)$ convergent since $a_n \neq o(n).$

### Euler-Maclaurin Sum Formula

**1.20  THEOREM.**  *If $f(x)$ is continuously differentiable on $[1,n]$ (for some fixed positive integer n) then*

$$\sum_{k=1}^{n} f(k) = \frac{1}{2} [f(1) + f(n)] + \int_1^n f(x) \, dx + \int_1^n (x - [x] - \frac{1}{2}) f^{(1)}(x) \, dx$$

*(where $[x]$ denotes the integer part of x, i.e. $[x] = k$ when $k \leqslant x < k+1$ and k is an integer).*

*Proof.*  Since

$$\int_1^n [x] \, f^{(1)}(x) dx = \sum_{k=1}^{n-1} \int_k^{k+1} [x] \, f^{(1)}(x) dx = \sum_{k=1}^{n-1} \int_k^{k+1} k f^{(1)}(x) dx$$

we have

$$\int_1^n (x - [x] - \frac{1}{2}) f^{(1)}(x) dx$$

$$= \sum_{k=1}^{n-1} \int_k^{k+1} (x - k - \frac{1}{2}) f^{(1)}(x) dx$$

$$= \sum_{k=1}^{n-1} \left\{ \int_k^{k+1} x f^{(1)}(x) dx - (k+\frac{1}{2}) (f(k+1) - f(k)) \right\}.$$

Integrating this last expression by parts yields

$$\int_1^n (x - [x] - \frac{1}{2}) f^{(1)}(x) dx$$

$$= \sum_{k=1}^{n-1} \{ (k+1) f(k+1) - kf(k) - \int_k^{k+1} f(x) dx - (k+\frac{1}{2}) (f(k+1) - f(k)) \}$$

$$= \sum_{k=1}^{n-1} \{ \frac{1}{2} f(k+1) + \frac{1}{2} f(k) - \int_k^{k+1} f(x) dx \}.$$

Therefore,

$$\int_1^n (x - [x] - \frac{1}{2}) f^{(1)}(x) dx$$

$$= - \frac{1}{2} \{ f(1) + f(n) \} + \sum_{k=1}^{n} f(x) - \int_1^n f(x) dx.$$

The conclusion of the theorem now follows directly.     $\triangle$

The summation formula given in the statement of Theorem 1.20 is known as the Euler–Maclaurin sum formula.

**1.21 EXAMPLE.** If $f(x) = 1/x$ (in Theorem 1.20) then

$$\sum_{k=1}^{n} \frac{1}{k} = \log n + \frac{1}{2}(1 + \frac{1}{n}) - I(n) \quad \text{where } I(n) = \int_{1}^{n} (x - [x] - \frac{1}{2}) \frac{dx}{x^2}. \quad (1.1)$$

Now, $|x - [x] - \frac{1}{2}| \leqslant \frac{1}{2}$ and thus $|I(n)| \leqslant \frac{1}{2}[1 - 1/n]$. Hence, from (1.1), we have

$$\sum_{k=1}^{n} \frac{1}{k} = \log n + O(1) \quad (1.2)$$

(note that $|I(n+p) - I(n)| = |\int_{n}^{n+p} (x - [x] - \frac{1}{2}) \frac{dx}{x^2}| \leqslant \frac{1}{2}(\frac{1}{n} - \frac{1}{n+p}) \leqslant \frac{1}{2n}$,

which is arbitrarily small for large $n$, thus $\lim_{n\to\infty} I(n)$ exists). Define

$$\gamma = -\lim_{n\to\infty} I(n) + \frac{1}{2} = \frac{1}{2} - \int_{1}^{\infty} (x - [x] - \frac{1}{2}) \frac{dx}{x^2}. \quad (1.3)$$

From (1.1) and (1.3) we have

$$\gamma = \lim_{n\to\infty} \{1 + \frac{1}{2} + \ldots + \frac{1}{n} - \log n\} \quad (1.4)$$

and, hence, we can rewrite (1.1) as

$$\sum_{k=1}^{n} \frac{1}{k} = \log n + \frac{1}{2} + \frac{1}{2n} - \int_{1}^{\infty} (x - [x] - \frac{1}{2}) \frac{dx}{x^2} + \int_{n}^{\infty} (x - [x] - \frac{1}{2}) \frac{dx}{x^2}$$

$$= \log n + \gamma + o(1).$$

The constant $\gamma$ is known as Euler's constant and is approximately equal to $0 \cdot 577$.

*Exercises.* (1) Show that the sequence $\{a_n\}_1^{\infty}$ (where $a_n = 1 + \frac{1}{2} + \ldots + (1/n) - \log n$) is monotone decreasing. Show that the sequence $\{b_n\}_1^{\infty}$ (where $b_n = 1 + \frac{1}{2} + \ldots + (1/n) - \log (n+1)$) is monotone increasing. Finally, show that $0 < a_n - b_n = o(1)$. Hint: Since $\log [(1 + \frac{1}{n})] = \int_{0}^{1/n} dt/(1+t)$ it follows that $1/n > \log [(1 + \frac{1}{n})]$.

Therefore,

$$a_{n+1} - a_n = \frac{1}{n+1} - \log(1 + \frac{1}{n}) < 0 \quad \text{and} \quad b_{n+1} - b_n = \frac{1}{n+1} - \log(1 + \frac{1}{n+1}) > 0.$$

Note that the sequence $\{a_n\}_1^{\infty}$ decreases to $\gamma$, the sequence $\{b_n\}_1^{\infty}$ increases to $\gamma$ and $b_n < \gamma < a_n$ for all $n \geqslant 1$.)

(2)   Show that

$$1 + \tfrac{1}{3} + \tfrac{1}{5} + \ldots + \frac{1}{2n-1} = \tfrac{1}{2}(\gamma + \log n) + \log 2 + o(1)$$

and that

$$1 - \tfrac{1}{2} + \tfrac{1}{3} - \tfrac{1}{4} + \ldots + \frac{1}{2n-1} - \frac{1}{2n} = \log 2 + o(1).$$

(Hint:

$$1 + \tfrac{1}{3} + \ldots + \frac{1}{2n-1} = 1 + \tfrac{1}{2} + \ldots + \frac{1}{2n} - \tfrac{1}{2}(1 + \tfrac{1}{2} + \ldots + \tfrac{1}{n})$$

$$= \log 2n + \gamma + o(1) - \tfrac{1}{2}(\log n + \gamma + o(1))$$

and

$$1 - \tfrac{1}{2} + \tfrac{1}{3} - \tfrac{1}{4} + \ldots + \frac{1}{2n-1} - \frac{1}{2n} = 1 + \tfrac{1}{2} + \ldots + \frac{1}{2n} - \tfrac{1}{2}(1 + \tfrac{1}{2} + \ldots + \tfrac{1}{n})$$

$$= \log 2n + \gamma + o(1) - (\log n + \gamma + o(1)).)$$

(3)   Prove that $\sum_{k=1}^{n} f(k) - \int_{1}^{n} f(x)\,dx$ tends to a limit (as $n \to \infty$) where $f(x)$ is a positive decreasing function of $x$ and continuous in $x$.
(Hint: $f(k) \geqslant \int_{k}^{k+1} f(x)\,dx \geqslant f(k+1)$ for all $k = 1, 2, \ldots, n-1$.   Adding these inequalities yields

$$A_n = \sum_{k=1}^{n} f(k) - \int_{1}^{n} f(x)\,dx = f(n) + \sum_{k=1}^{n-1} f(k) - \int_{1}^{n} f(x)\,dx \geqslant f(n) \geqslant 0.$$

Also, $A_n - A_{n-1} = f(n) - \int_{n-1}^{n} f(x)\,dx \leqslant 0$ and $A_1 = f(1) > 0$.   Hence, the sequence $\{A_n\}_1^{\infty}$ is convergent.   Note that $f(n) \leqslant A_n \leqslant f(1)$.)

(4)   Let $f(x)$ be differentiable for $x \geqslant 1$ and suppose that $f^{(1)}(x)$ tends monotonically to zero as $x \to \infty$,   Prove that

$$\lim_{n \to \infty} (\sum_{k=1}^{n} f(k) - \tfrac{1}{2}(f(1) + f(n)) - \int_{1}^{n} f(x)\,dx)$$

exists.   Consider the special case when $f(x) = -\log x$ and show that $\log n! = n \log n - n + \tfrac{1}{2} \log n + O(1)$.   The right-hand side of this last equality gives us the first three terms in Stirling's formula, i.e.

$$\log n! = n \log n - n + \tfrac{1}{2} \log n + \tfrac{1}{2} \log(2\pi) + o(1).$$

(5)   Let $f(x)$ be differentiable for $x \geqslant 1$ and suppose that $f^{(1)}(x)$ increases to $\infty$ as $x \to \infty$.   Show that

$$\sum_{k=1}^{n} f(k) = \tfrac{1}{2}(f(1) + f(n)) + \int_{1}^{n} f(x)\,dx + O(f^{(1)}(n)).$$

Consider the special case when $f(x) = x^p$ for $p > 1$.

(6)   If the sequence $\{b_n\}_1^{\infty}$ increases to $\infty$ then prove that

$$\lim \sup\nolimits_{n \to \infty} \frac{a_{n+1} - a_n}{b_{n+1} - b_n} \;\geqslant\; \lim \sup\nolimits_{n \to \infty} \frac{a_n}{b_n} \;\geqslant\; \lim \inf\nolimits_{n \to \infty} \frac{a_n}{b_n}$$

$$\geqslant \lim \inf\nolimits_{n \to \infty} \frac{a_{n+1} - a_n}{b_{n+1} - b_n} .$$

Here we do not require $a_n = o(1)$ as we did in Example 1.17.

(7)  Prove:  If the series $\sum_{k=0}^{\infty} a_k$ is convergent then  $T_n = a_1 + 2a_2$

$+ \ldots + na_n = o(n)$.  (*Hint:* Let $S_n = \sum_{k=0}^{n} a_k$ and $\sigma_n = \dfrac{1}{n+1} \sum_{j=0}^{n} S_j$.

We have $(n+1) S_n - (n+1) \sigma_n = T_n$.  Now, $\{S_n\}_0^{\infty}$ converges to $S$ and, hence, $\{\sigma_n\}_0^{\infty}$ converges to $S$.  Therefore $T_n = o(n)$.  Alternatively, we can prove this result by taking $b_n = n+1$, $a_n = S_0 + \ldots + S_n$ in (6). This gives us

$$\lim\nolimits_{n \to \infty} S_{n+1} = \lim\nolimits_{n \to \infty} \frac{S_0 + \ldots + S_n}{n+1}$$

i.e.        $\lim\nolimits_{n \to \infty} S_n = \lim\nolimits_{n \to \infty} \sigma_n$  and so  $T_n = o(n)$.)

(8)  Prove:  If the sequence of partial sums $\{S_n\}_0^{\infty}$ of $\sum_{k=0}^{\infty} a_k$ tends to $\infty$ then the sequence of $(C,1)$ partial sums $\{\sigma_n\}_0^{\infty}$ also tends to $\infty$. (*Hint:* Take $b_n = n+1$, $a_n = S_0 + \ldots + S_n$ and, then, we obtain

$$\lim\nolimits_{n \to \infty} \frac{1}{n+1} \sum_{k=0}^{n} S_k = \lim\nolimits_{n \to \infty} S_{n+1} = \infty .)$$

(9)  Let $b(x) = \sum_{k=0}^{\infty} b_k x^k$ where $b_k \geqslant 0$ for all $k = 0, 1, \ldots$ .  Suppose that $\sum_{k=0}^{\infty} b_k x^k$ converges for $|x| < 1$ and diverges at $x = 1$.  If $a(x) = \sum_{k=0}^{\infty} a_k x^k$ where $a_n \sim cb_n$, $(n \to \infty)$ and $c$ is a real constant $(c \neq 0)$ then $a(x) \sim cb(x)$, $(x \to 1)$.  (*Hint:* By our hypothesis, the series representing $a(x)$ has radius of convergence 1.  We have, given $\varepsilon > 0$, that there exists a positive integer $N = N(\varepsilon)$ such that $|a_n - cb_n| < \varepsilon b_n$ for $n \geqslant N$.  So

$$|a(x) - cb(x)| = \left| \sum_{k=0}^{\infty} (a_k - cb_k)x^k \right| \leqslant \Sigma_1 + \Sigma_2$$

where        $\Sigma_1 = \left| \sum_{k=0}^{N} (a_k - cb_k)x^k \right| \leqslant \sum_{k=0}^{N} |a_k - cb_k|$

and        $\Sigma_2 = \left| \sum_{k=N+1}^{\infty} (a_k - cb_k)x^k \right| \leqslant c \sum_{k=N+1}^{\infty} b_k x^k < \varepsilon b(x).$

Fix $N$ and choose $\delta$ so that $\Sigma_1 < \varepsilon b(x)$ for $x > 1-\delta$.    Thus for $1-\delta < x < 1$ we have $|a(x) - cb(x)| < 2\varepsilon b(x)$.)

*Remark:*   Our argument shows that if $a_n = o(b_n)$ then $a(x) = o(b(x))$, $(x \to 1)$.  A similar remark applies to Exercises (10) and (11).

(10)   Let $b(x) = \sum\limits_{k=0}^{\infty} b_k x^k$ be convergent for $|x| < 1$.  Suppose that $B_n = \sum\limits_{k=0}^{n} b_k \geqslant 0$ for $n \geqslant 1$ and that the sequence $\{B_n\}_1^{\infty}$ tends to $\infty$ as $n \to \infty$.  If $a(x) = \sum\limits_{k=0}^{\infty} a_k x^k$ and $A_n = \sum\limits_{k=0}^{n} a_k \sim cB_n$, $(n \to \infty)$ where $c$ is a real constant then $a(x) \sim cb(x)$, $(x \to 1)$.        (Note that $a(x) = (1-x) \sum\limits_{k=0}^{\infty} A_k x^k$ and $b(x) = (1-x) \sum\limits_{k=0}^{\infty} B_k x^k$.)

(11)   Prove:  If $c \neq 0$ and $\sum\limits_{k=0}^{n} a_k \sim cn$, $(n \to \infty)$, then $\sum\limits_{k=0}^{\infty} a_k x^k \sim c/(1-x)$, $(x \to 1)$.  (*Hint:*  $b(x) = \sum\limits_{k=0}^{\infty} kx^k = (1-x)x(1-x)^{-2}$.)

(12)   Prove:  If $\theta$ is real then

(i)    $|\sin \theta| \leqslant |\theta|$,

(ii)   $1 - \cos \theta \leqslant \frac{1}{2}\theta^2$,

(iii)  $\sin \theta \geqslant \dfrac{2\theta}{\pi}$  $(0 \leqslant \theta \leqslant \frac{1}{2}\pi)$, and

(iv)   $1 - \cos \theta \geqslant \dfrac{2}{\pi^2}\theta^2$  $(0 \leqslant \theta \leqslant \pi)$.

## Abel's Inequality

The following result is due to Abel and is referred to as Abel's inequality.

1.22   THEOREM.   *If $S_k = \sum\limits_{j=1}^{k} a_j$, $a_j$ real, $|S_k| \leqslant A$ for $k = 1, 2, \ldots, n$, and $q_1 \geqslant q_2 \geqslant \ldots \geqslant q_n > 0$ then $\left|\sum\limits_{k=1}^{n} a_k q_k\right| \leqslant Aq_1$.*

*Proof:*   Write $S_0 = 0$.   Thus

$$\left|\sum_{k=1}^{n} a_k q_k\right| = \left|\sum_{k=1}^{n} (S_k - S_{k-1}) q_k\right| = \left|\sum_{k=0}^{n-1} S_k q_k - \sum_{k=0}^{n-1} S_k q_{k+1} + S_n q_n\right|$$

$$= \left|\sum_{k=1}^{n-1} S_k (q_k - q_{k+1}) + S_n q_n\right|$$

$$\leqslant A \left\{ \sum_{k=1}^{n-1} (q_k - q_{k+1}) + q_n \right\} = Aq_1. \qquad \triangle$$

**1.23** EXAMPLE. The series

$$\sum_{n=1}^{\infty} \frac{\sin nx}{n}$$

(i)  is convergent for all real $x$ and   (ii)  is uniformly convergent in any closed interval not containing multiples of $2\pi$.   Write

$$D_n(x) = \tfrac{1}{2} + \ldots + \cos nx$$

and

$$\widetilde{D}_n(x) = \sin x + \ldots + \sin nx$$

thus

$$2D_n(x) \sin \tfrac{1}{2}x = \sin \tfrac{1}{2}x + \sum_{k=1}^{n} 2 \sin \tfrac{1}{2}x \cos kx$$

$$= \sin \tfrac{1}{2}x + \sum_{k=1}^{n} \left\{ \sin(k + \tfrac{1}{2})x - \sin(k - \tfrac{1}{2})x \right\}$$

$$= \sin(n + \tfrac{1}{2})x.$$

Hence

$$D_n(x) = \begin{cases} \dfrac{\sin(n + \tfrac{1}{2})x}{2 \sin \tfrac{1}{2}x} & \text{if } x \not\equiv 0 \pmod{2\pi} \\[4mm] n + \tfrac{1}{2} & \text{if } x \equiv 0 \pmod{2\pi}. \end{cases}$$

Similarly

$$\widetilde{D}_n(x) = \begin{cases} \dfrac{\cos \tfrac{1}{2}x - \cos(n + \tfrac{1}{2})x}{2 \sin \tfrac{1}{2}x} & \text{if } x \not\equiv 0 \pmod{2\pi} \\[4mm] 0 & \text{if } x \equiv 0 \pmod{2\pi}. \end{cases}$$

Hence, for $x \not\equiv 0 \pmod{2\pi}$,

$$\left| \sum_{k=m+1}^{n} \sin kx \right| = \left| \sum_{k=1}^{n} \sin kx - \sum_{k=1}^{m} \sin kx \right|$$

$$\leqslant \frac{1}{\left| \sin \tfrac{1}{2}x \right|}$$

and

$$\left| \sum_{k=m+1}^{n} \cos kx \right| \leqslant \frac{1}{\left| \sin \tfrac{1}{2}x \right|}.$$

Let $q_1 = 1/(m+1)$, ... , $q_{n-m} = 1/n$ and $a_1 = \sin(m+1)x$, ... , $a_{n-m} = \sin nx$.   Thus Abel's inequality gives

$$\left| \sum_{k=m+1}^{n} \frac{\sin kx}{k} \right| \leqslant \frac{1}{m+1} \; \frac{1}{\left| \sin \frac{1}{2}x \right|}.$$

This proves the convergence for $x$ a non-multiple of $2\pi$.   If $x$ is a multiple of $2\pi$ then the series converges to 0 trivially.

Note that for $2k\pi + \delta \leqslant x \leqslant 2(k+1)\pi - \delta$, $0 < \delta < \pi$, $k$ integer or zero, we have

$$\left| \sum_{k=m+1}^{n} \frac{\sin kx}{k} \right| \leqslant \frac{1}{m+1} \; \frac{1}{\sin \frac{1}{2}\delta}.$$

This proves part (ii).


*Exercises.*   Suppose $\{a_n\}_1^\infty$ is a decreasing sequence of positive numbers tending to zero.     Prove:   the series

$$\sum_{n=1}^{\infty} a_n \sin nx$$

(i)  is convergent for all real $x$;   (ii)  is uniformly convergent in any closed interval not containing multiples of $2\pi$;   (iii)  the series

$$\sum_{n=1}^{\infty} a_n \cos nx$$

is convergent for all real $x$ other than multiples of $2\pi$ and is uniformly convergent in any interval not containing multiples of $2\pi$.

(ii)   Prove that the series

$$\sum_{n=1}^{\infty} \frac{\sin nx}{n}$$

is not absolutely convergent for any real $x$ except when $x$ is a multiple of $\pi$ and the series

$$\sum_{x=1}^{\infty} \frac{\cos nx}{n}$$

is not absolutely convergent for any (real) $x$.

(*Hint:*

$$\left| \frac{\cos nx}{n} \right| \geqslant \frac{1}{n} \cos^2 nx = \frac{1}{2n}(1 + \cos 2nx)$$

and

$$\left| \frac{\sin nx}{n} \right| \geqslant \frac{1}{n} \sin^2 nx = \frac{1}{2n}(1 - \cos 2nx).)$$

# CHAPTER 2

## General Summability Theory

We saw, in Chapter 1, two different ways to define convergence (the Abel method and the Cesàro method of order 1 ((C, 1) method) and, in the case of the (C,1) method, the example of convergence of the Cauchy product of two series leads one to believe that other forms of convergence are, indeed, legitimate to examine for reasons other than just that of being a mathematical exercise. In fact, in Chapters 5 and 6, we examine some other applications of summability transforms which should help convince the reader of the usability and desirability of the study of summability.

In this chapter we undertake a study of some of the notions of summability in general. We follow this chapter (in Chapter 3) with an examination of particular well-known transforms.

### Basic Definitions and Concepts

Let $A = (a_{n,k})$ be the 'infinite matrix

$$
\begin{bmatrix}
a_{0,0} & a_{0,1} & a_{0,2} & \cdots & a_{0,k} & \cdots \\
a_{1,0} & a_{1,1} & a_{1,2} & \cdots & a_{1,k} & \cdots \\
\cdot & \cdot & \cdot & & \cdot & \\
\cdot & \cdot & \cdot & & \cdot & \\
\cdot & \cdot & \cdot & & \cdot & \\
a_{n,0} & a_{n,1} & a_{n,2} & \cdots & a_{n,k} & \cdots \\
\cdot & \cdot & \cdot & & \cdot & \\
\cdot & \cdot & \cdot & & \cdot & \\
\cdot & \cdot & \cdot & & \cdot & \\
\end{bmatrix}
$$

where $a_{n,k}$ is a complex number for all $n = 0, 1, \ldots$ and $k = 0, 1, \ldots$ .

21

**2.1 DEFINITION.** *Let* $\{z_n\}_0^\infty$ *be a sequence of complex numbers and let* $A = (a_{n,k})$ *be an infinite matrix. If*

$$\sigma_n = \sum_{k=0}^{\infty} a_{n,k} z_k \text{ converges for each } n = 0, 1, \ldots$$

*then the sequence* $\{\sigma_n\}_0^\infty$ *is called the A-transform of the sequence* $\{z_n\}_0^\infty$.

If we extend the idea of matrix multiplication to infinite matrices we can write

$$
\begin{bmatrix}
a_{0,0} & a_{0,1} & \cdots & a_{0,k} & \cdots \\
a_{1,0} & a_{1,1} & \cdots & a_{1,k} & \cdots \\
\cdot & \cdot & & \cdot & \\
\cdot & \cdot & & \cdot & \\
\cdot & \cdot & & \cdot & \\
a_{n,0} & a_{n,1} & \cdots & a_{n,k} & \cdots \\
\cdot & \cdot & & \cdot &
\end{bmatrix}
\begin{bmatrix}
z_0 \\
z_1 \\
\cdot \\
\cdot \\
\cdot \\
z_k \\
\cdot
\end{bmatrix}
=
\begin{bmatrix}
\sum_{k=0}^{\infty} a_{0,k} z_k \\
\sum_{k=0}^{\infty} a_{1,k} z_k \\
\cdot \\
\cdot \\
\cdot \\
\sum_{k=0}^{\infty} a_{n,k} z_k \\
\cdot
\end{bmatrix}
=
\begin{bmatrix}
\sigma_0 \\
\sigma_1 \\
\cdot \\
\cdot \\
\cdot \\
\sigma_n \\
\cdot
\end{bmatrix}.
$$

Using this idea let

$$a_{n,k} = \begin{cases} 0 & \text{if } k > n \\[2mm] \dfrac{1}{n+1} & \text{if } k \leqslant n \end{cases}$$

and let $\{z_n\}_0^\infty$ be a sequence of complex numbers. The $A = (a_{n,k})$-transform of the sequence $\{a_n\}_0^\infty$ is the sequence $\{\sigma_n\}_0^\infty$ where

$$\sigma_n = \sum_{k=0}^{\infty} a_{n,k} z_k = \sum_{k=0}^{n} \left(\frac{1}{n+1}\right) z_k = \frac{1}{n+1} \sum_{k=0}^{n} z_k.$$

Therefore, if $\{\sigma_n\}_0^\infty$ converges then we have that the sequence $\{z_n\}_0^\infty$ is $(C,1)$ convergent (which we studied in Chapter 1). This particular transform is called the Cesàro (or $(C,1)$) transform.

In general, given a matrix (transform) $A = (a_{n,k})$, we want to be able to answer the question 'If $\{z_n\}_0^\infty$ is any sequence of complex numbers which converges (say to $z$) when does the $A$-transform of $\{z_n\}_0^\infty$ converge (and, in particular, converges to the same limit $z$)?'. Notice that the term 'transform' is used to denote the matrix $A$ while the term '$A$-transform' denotes the resulting sequence when

*A* is applied to a sequence as described.

2.2  DEFINITION.  *Let* $A = (a_{n,k})$ *be an infinite matrix of complex numbers.*

(i)  *If the A-transform of any convergent sequence of complex numbers exists and converges then A is conservative.*

(ii)  *If A is conservative and preserves limits, i.e.* $\lim_{n\to\infty} z_n = z$ *implies* $\lim_{n\to\infty} \sigma_n = z$ *where* $\{\sigma_n\}_0^\infty$ *is the A-transform of the convergent sequence* $\{z_n\}_0^\infty$, *then A is regular.*

(iii)  *If the A-transform of the sequence* $\{z_n\}_0^\infty$ *converges to y then* $\{z_n\}_0^\infty$ *is A-summable to y.*

Referring back to the (C,1)-transform we see that this matrix is regular (see Theorem 1.10).

## The Silverman-Toeplitz Theorem

Our primary interest, now, is to find out when a given transform $A = (a_{n,k})$ is regular. We can, of course, directly apply Definition 2.2 but, as one might guess, this can frequently be most difficult. In the study of the Calculus we can recall defining what is meant by taking the derivative of a function. We quickly discovered that it would be to our advantage to develop some general theorems concerning the differentiation of various types of functions (such as the differentiation of sums, products, and quotients of functions, polynominals, trigonometric functions, etc.) instead of applying, directly, the definition of differentiation to each particular function which we encountered. Similarly, we would like to be able to determine if a matrix $A = (a_{n,k})$ is regular merely by examining the entries, $a_{n,k}$, of the matrix. The main theorem which we will prove (known as the Silverman–Toeplitz theorem) states

2.3  THEOREM.  *A matrix* $A = (a_{n,k})$ *is regular if and only if*

(i)  $\lim_{n\to\infty} a_{n,k} = 0$  *for each* $k = 0, 1, \ldots$ ,

(ii)  $\lim_{n\to\infty} \sum_{k=0}^{\infty} a_{n,k} = 1$,  *and*

(iii)  $\sup_n \{ \sum_{k=0}^{\infty} |a_{n,k}| \} \leqslant M < \infty$  *for some* $M > 0$.

*Proof.* We will first prove that conditions (i), (ii), and (iii) imply that $A = (a_{n,k})$ is regular.

Let $\{z_k\}_0^\infty$ be a sequence of complex numbers converging to $z$. Thus, there exists $T > 0$ such that $|z_k| \leqslant T$ for all $k = 0, 1, \ldots$ .

By condition (iii) we have that $\sum\limits_{k=0}^{\infty} |a_{n,k}|$ converges for each fixed $n$. Therefore, $\sum\limits_{k=0}^{\infty} |a_{n,k}| T$ converges which implies that $\sum\limits_{k=0}^{\infty} |a_{n,k}| |z_k|$ converges (we should note that we only used the boundedness of $\{z_k\}_0^{\infty}$ to prove this fact).

Next, assuming that $\{z_n\}_0^{\infty}$ converges to $z$, we want to prove that $\{\sigma_n\}_0^{\infty}$ converges to $z$ also (where $\sigma_n = \sum\limits_{k=0}^{\infty} a_{n,k} z_k$). Consider the case when $\{z_n\}_0^{\infty}$ converges to $z = 0$. Let $\varepsilon > 0$ be given. We want find a positive integer $N(\varepsilon)$ such that if $n \geqslant N(\varepsilon)$ then $|\sigma_n - z| = |\sigma_n| < \varepsilon$, i.e. $|\sum\limits_{k=0}^{\infty} a_{n,k} z_k| < \varepsilon$. Since $\{z_n\}_0^{\infty}$ converges to 0 there exists a non-negative integer $N_1(\varepsilon) = N_1$ such that if $n \geqslant N_1$ then $|z_n| \leqslant \varepsilon/(2M+1)$ (where $\sup_n \{\sum\limits_{k=0}^{\infty} |a_{n,k}|\} \leqslant M < \infty$) and $|z_{N_1 - 1}| \geqslant \varepsilon/(2M+1) > 0$ (if $N_1 = 0$ then we have $|z_n| < \varepsilon/(2M+1)$ for all $n = 0, 1, \ldots$). Now, if $N_1 = 0$ then

$$|\sigma_n| = |\sum_{k=0}^{\infty} a_{n,k} z_k| \leqslant \sum_{k=0}^{\infty} |a_{n,k}| |z_k| < \frac{\varepsilon}{2M+1} \sum_{k=0}^{\infty} |a_{n,k}| \leqslant \frac{\varepsilon}{2M+1} (M) < \varepsilon.$$

Now, suppose that $N_1 \neq 0$. We have

$$|\sigma_n| = |\sum_{k=0}^{\infty} a_{n,k} z_k| \leqslant |\sum_{k=0}^{N_1-1} a_{n,k} z_k| + \sum_{k=N_1}^{\infty} |a_{n,k}| |z_k|. \qquad (2.1)$$

Consider

$$\sum_{k=N_1}^{\infty} |a_{n,k}| |z_k| < \frac{\varepsilon}{2M+1} \sum_{k=N_1}^{\infty} |a_{n,k}|$$

$$\leqslant \frac{\varepsilon}{2M+1} \sum_{k=0}^{\infty} |a_{n,k}| \leqslant \frac{\varepsilon}{2M+1} (M) < \frac{1}{2}\varepsilon \qquad (2.2)$$

(by condition (iii)). Let $T = \max \{|z_0|, \ldots, |z_{N_1-1}|\}$ (which is positive since $N_1 > 0$). There exist positive integers $N_2(0, \varepsilon), \ldots, N_2(N_1 - 1, \varepsilon)$ such that if $n \geqslant N_2(k, \varepsilon)$ then $|a_{n,k}| < \varepsilon/2TN_1$ for each $k = 0, \ldots, N_1 - 1$ (by condition (i)). Let $N(\varepsilon) = \max \{N_2(0, \varepsilon), \ldots, N_2(N_1 - 1, \varepsilon)\}$. If $n \geqslant N(\varepsilon)$ then

$$|\sum_{k=0}^{N_1-1} a_{n,k} z_k| \leqslant \sum_{k=0}^{N_1-1} |a_{n,k}| |z_k| < N_1 \frac{\varepsilon}{2TN_1} T = \frac{1}{2}\varepsilon. \qquad (2.3)$$

Therefore, if $n \geqslant N(\varepsilon)$ then, by (2.1), (2.2), and (2.3) we have

$$|\sigma_n| \leqslant |\sum_{k=0}^{N_1-1} a_{n,k} z_k| + \sum_{k=N_1}^{\infty} |a_{n,k}| |z_k| < \frac{1}{2}\varepsilon + \frac{1}{2}\varepsilon = \varepsilon,$$

thus the sequence $\{\sigma_n\}_0^{\infty}$ converges to 0.

Now, we remove the condition that $\{z_n\}_0^{\infty}$ converges to 0. Suppose that $\{z_n\}_0^{\infty}$ converges to $z$. Define the sequence $\{w_n\}_0^{\infty}$ by $w_n = z_n - z$ for each $n = 0, 1, \ldots$ . Since $\{w_n\}_0^{\infty}$ converges to 0 we have that $\lim_{n\to\infty} \sum_{k=0}^{\infty} a_{n,k} w_k = 0$, i.e.

$$0 = \lim_{n\to\infty} \sum_{k=0}^{\infty} a_{n,k}(z_k - z) = \lim_{n\to\infty} (\sum_{k=0}^{\infty} a_{n,k} z_k - \sum_{k=0}^{\infty} a_{n,k} z)$$

$$= \lim_{n\to\infty} \sum_{k=0}^{\infty} a_{n,k} z_k - z \lim_{n\to\infty} \sum_{k=0}^{\infty} a_{n,k} = \lim_{n\to\infty} \sum_{k=0}^{\infty} a_{n,k} z_k - z$$

(by condition (ii)). Therefore, the $A$-transform of $\{z_n\}_0^{\infty}$ converges to $z$ and, hence, $A = (a_{n,k})$ is regular.

We now want to prove that the regularity of $A = (a_{n,k})$ implies that the conditions (i), (ii), and (iii) hold.

Let $k$ be fixed. Define the sequence $\{z_n\}_0^{\infty}$ by

$$z_n = \begin{cases} 0 & \text{if } n \neq k \\ 1 & \text{if } n = k. \end{cases}$$

Clearly, $\{z_n\}_0^{\infty}$ converges to 0. Since $A = (a_{n,k})$ is regular we have that

$$\lim_{n\to\infty} \sigma_n = \lim_{n\to\infty} \sum_{j=0}^{\infty} a_{n,j} z_j = \lim_{n\to\infty} a_{n,k} = 0.$$

This proves condition (i).

Define the sequence $\{w_n\}_0^{\infty}$ by $w_n = 1$ for each $n = 0, 1, \ldots$ . Since $\{w_n\}_0^{\infty}$ converges to 1 and $A = (a_{n,k})$ is regular we have that $\lim_{n\to\infty} \sigma_n = \lim_{n\to\infty} \sum_{k=0}^{\infty} a_{n,k} w_k = \lim_{n\to\infty} \sum_{k=0}^{\infty} a_{n,k} = 1$. So condition (ii) holds.

To demonstrate that condition (iii) holds is a more complicated task.

First, we show that $\sum_{k=0}^{\infty} |a_{n,k}|$ converges for each $n$. Suppose, to the contrary, that there exists $N$ such that $\sum_{k=0}^{\infty} |a_{N,k}|$ diverges (i.e. $\sum_{k=0}^{\infty} |a_{N,k}| = \infty$). Therefore, there exist integers $0 = K_0 < K_1 < \ldots < K_j < \ldots$ such that $\sum_{k=K_{j-1}}^{K_j-1} |a_{N,k}| > 1$ for each $j = 1, 2, \ldots$ . Define the sequence $\{z_k\}_0^{\infty}$ by

$$z_k = \begin{cases} 0 & \text{if } k = 0 \text{ or } a_{N,k} = 0 \\ \\ \dfrac{|a_{N,k}|}{j\,a_{N,k}} & \text{if } a_{N,k} \neq 0 \text{ and } K_{j-1} \leqslant k < K_j \quad (\text{for } j = 1, 2, \ldots). \end{cases}$$

Since $\{z_k\}_0^\infty$ converges (to 0) and $A = (a_{n,k})$ is regular, we have that $\sum\limits_{k=0}^\infty a_{n,k} z_k$ converges for each $n = 0, 1, \ldots$, in particular for $n = N$. However,

$$\sum_{k=0}^\infty a_{N,k} z_k = \sum_{j=1}^\infty \sum_{k=K_{j-1}}^{K_j - 1} \frac{1}{j} |a_{N,k}|$$

$$= \sum_{j=1}^\infty \frac{1}{j} \sum_{k=K_{j-1}}^{K_j - 1} |a_{N,k}| \geqslant \sum_{j=1}^\infty \frac{1}{j}$$

which diverges. This gives us a contradiction and, hence, $\sum\limits_{k=0}^\infty |a_{n,k}|$ converges for each $n = 0, 1, \ldots$ .

Now, to show that the sequence $\{\sum\limits_{k=0}^\infty |a_{n,k}|\}_0^\infty$ is bounded we assume that $\sup_n \{\sum\limits_{k=0}^\infty |a_{n,k}|\} = \infty$ and obtain a contradiction.

Construct the two sequences of integers $\{m_k\}_0^\infty$ and $\{n_k\}_0^\infty$ in the following manner.

Choose $m_0 = 0$ and choose

$$n_0 = \min\{n: n \geqslant 1 \quad \text{and} \quad \sum_{k=n+1}^\infty |a_{m_0,k}| < 1\}$$

(since $\sum\limits_{k=0}^\infty |a_{m_0,k}|$ converges $n_0$ exists). In general, assume that we have chosen $m_0, \ldots, m_{j-1}$ and $n_0, \ldots, n_{j-1}$ $(j \geqslant 1)$. Choose

$$m_j = \min\{m: m > m_{j-1}, \sum_{k=0}^\infty |a_{m\,k}| > j^2 + 2j + 2 \quad \text{and} \quad \sum_{k=0}^{n_{j-1}} |a_{m,k}| < 1\}.$$

The integer $m_j$ exists since

   (i)  we can choose $M_1 > m_{j-1}$ such that $|a_{M,k}| < 1/(n_{j-1} + 1)$ for $k = 0, \ldots, n_{j-1}$ and all $M \geqslant M_1$ (using the fact that $\lim_{n \to \infty} a_{n,k} = 0$ for fixed $k$), and

   (ii) we can choose $M_2 \geqslant M_1$ such that $\sum\limits_{k=0}^\infty |a_{M_2,k}| > j^2 + 2j + 2$ (using the assumption that $\sup_n \{\sum\limits_{k=0}^\infty |a_{n,k}|\} = \infty$).

Choose $n_j = \min \{n: n > n_{j-1} \text{ and } \sum\limits_{k=n+1}^\infty |a_{m_j,k}| < 1\}$.          The integer $n_j$ exists because $\sum\limits_{k=0}^{\infty} |a_{m_j,k}|$ converges.

Define the sequence $\{z_k\}_0^{\infty}$ by

$$z_k = \begin{cases} \dfrac{|a(m_j,k)|}{ja(m_j,k)} & \text{if } n_{j-1} < k \leqslant n_j \ (j = 1, 2, \dots) \text{ and } a_{m_j,k} \neq 0 \\ \\ 0 & \text{otherwise.} \end{cases}$$

The sequence $\{z_k\}_0^{\infty}$ converges (to 0) and, since $A$ is regular, the sequence $\{\sigma_n\}_0^{\infty}$ must also converge where $\sigma_n = \sum\limits_{k=0}^{\infty} a_{n,k} z_k$. However,

$$|\sigma(m_j)| = \Big| \sum_{k=0}^{\infty} a_{m_j,k} z_k \Big| = \Big| \sum_{k=0}^{n_{j-1}} a_{m_j,k} z_k + \sum_{k=n_{j-1}+1}^{n_j} a_{m_j,k} z_k$$

$$+ \sum_{k=n_j+1}^{\infty} a_{m_j,k} z_k \Big|$$

$$\geqslant \Big| \sum_{k=n_{j-1}+1}^{n_j} a_{m_j,k} z_k \Big| - \sum_{k=0}^{n_{j-1}} |a_{m_j,k}| |z_k|$$

$$- \sum_{k=n_j+1}^{\infty} |a_{m_j,k}| |z_k|.$$

Therefore,

$$|\sigma(m_j)| \geqslant \frac{1}{j} \sum_{k=n_{j-1}+1}^{n_j} |a_{m_j,k}| - 1 - 1$$

$$= \frac{1}{j} \sum_{k=0}^{\infty} |a_{m_j,k}| - \sum_{k=0}^{n_{j-1}} |a_{m_j,k}| - \sum_{k=n_j+1}^{\infty} |a_{m_j,k}| - 2$$

$$\geqslant \frac{1}{j} ((j^2 + 2j + 2) - 1 - 1) - 1 = \frac{1}{j} (j^2 + 2j) - 2 = j.$$

Therefore, $\{\sigma_n\}_0^{\infty}$ diverges since a subsequence, $\{\sigma(m_j)\}_0^{\infty}$, of $\{\sigma_n\}_0^{\infty}$ is unbounded. This gives us our contradiction. Therefore, $\sup_n \{ \sum\limits_{k=0}^{\infty} |a_{n,k}| \} \leqslant M < \infty$ for some $M > 0$ and this completes the proof of the theorem. $\triangle$

The major difficulty in proving Theorem 2.3 was to demonstrate that regularity of $A = (a_{n,k})$ implies $\sup_n \{ \sum\limits_{k=0}^{\infty} |a_{n,k}| \} \leqslant M < \infty$ for some $M > 0$. A much shorter proof can be given of this fact if one has, at his disposal, the use of the uniform boundedness principle (from functional analysis). This proof is given in the Appendix.

Let $A = (a_{n,k})$ and interpret $A^m = (a^m_{n,k})$ ($m$ a positive integer) to be

$$(a^m_{n,k}) = (a^1_{n,k})\,(a^{m-1}_{n,k}) \quad \text{where } a^1_{n,k} = a_{n,k}$$

provided this product exists.     This product is the usual matrix multiplication.

We say that a matrix is row finite provided it has at most a finite number of non-zero entries in each row.     Combining these two concepts with the concept of regularity, we have

2.4  THEOREM.  *If* $A = (a_{n,k})$ *is such that* $a_{n,k} = 0$ *for* $k > n$ *and* $A$ *is regular then* $A^m$ *is regular (m any positive integer).*

*Proof.*  We have that $A^1 = A = (a^1_{n,k})$ is regular.  Assume that $A^k$ is regular for $k = 1, \ldots, m-1$ ($m \geqslant 2$).  We want to prove that $A^m$ is regular.  Let $\{z_k\}^\infty_0$ be a sequence of complex numbers converging to $z$.  Let $\{\sigma_n\}^\infty_0$ be the $A^m = (a^m_{n,k})$ transform of $\{z_k\}^\infty_0$, i.e.

$$\sigma_n = \sum_{k=0}^{\infty} a^m_{n,k} z_k = \sum_{k=0}^{n} a^m_{n,k} z_k .$$

Let $\{t_n\}^\infty_0$ be the $A^{m-1} = (a^{m-1}_{n,k})$ transform of $\{z_k\}^\infty_0$,

i.e. $$t_n = \sum_{k=0}^{\infty} a^{m-1}_{n,k} z_k = \sum_{k=0}^{n} a^{m-1}_{n,k} z_k .$$

The sequence $\{t_n\}^\infty_0$ converges to $z$ since $A^{m-1}$ is regular.  We have that

$$\sigma_n = \sum_{k=0}^{n} a^m_{n,k} z_k = \sum_{k=0}^{n} \left( \sum_{h=k}^{n} a^1_{n,h}\, a^{m-1}_{h,k} \right) z_k$$

$$= \sum_{h=0}^{n} a^1_{n,h} \left( \sum_{k=0}^{h} a^{m-1}_{h,k} z_k \right) = \sum_{h=0}^{n} a^1_{n,h}\, t_n .$$

Since $A^1 = (a^1_{n,k})$ is regular and $\sigma_n$ is merely the $A^1$ transform of the sequence $\{t_n\}^\infty_0$ (which converges to $z$) we have that the sequence $\{\sigma_n\}^\infty_0$ converges to $z$, i.e. $A^m$ is regular.   $\triangle$

*Exercise.*  Prove:  If $A$ is row-finite and regular then $A^m$ is regular ($m$ any positive integer).

We now examine the behaviour, in general, of regular trans-formations with respect to summing divergent series.  We answer questions such as, 'Is there a regular transformation which sums all bounded divergent sequences?' and, since the answer to the above is in the negative (see Theorem 2.5), 'Is there a bounded divergent sequence which cannot be summed by any regular transformation?' This answer (see Theorem 2.6) is also in the negative.

2.5 THEOREM. *If $A = (a_{n,k})$ is a regular matrix then there exists a bounded divergent sequence $\{z_n\}_0^\infty$ such that the A-transform of $\{z_n\}_0^\infty$ diverges.*

*Proof.* Assume that $A = (a_{n,k})$ is regular. Therefore, by Theorem 2.3, we have

(i)   $\lim_{n\to\infty} a_{n,k} = 0$   for each $k = 0, 1, \ldots$,

(ii)   $\lim_{n\to\infty} \sum_{k=0}^\infty a_{n,k} = 1$,   and

(iii)   $\sup_n \left\{ \sum_{k=0}^\infty |a_{n,k}| \right\} \leqslant M < \infty$ for some $M > 0$.

Construct two sequences of positive integers $\{m_k\}_0^\infty$ and $\{n_k\}_0^\infty$ in the following manner (we will begin to use the notation $a(n,k)$ for $a_{n,k}$ now in order to avoid obvious printing difficulties).

There exists a positive integer $N$ such that $\left| 1 - \sum_{k=0}^\infty a(n,k) \right| < \frac{1}{8}$ for $n \geqslant N$, in particular, if $n \geqslant N$ then $\left| \sum_{k=0}^\infty a(n,k) \right| > \frac{7}{8}$ (by condition (ii)). Choose $m_0 = N$. Choose $n_0$ (a positive integer) such that $\sum_{k=n_0}^\infty |a(m_0,k)| < \frac{1}{8}$ (by condition (iii)).

In general, assume that we have chosen $m_0, \ldots, m_{j-1}$ and $n_0, \ldots, n_{j-1}$ ($j \geqslant 1$). Choose $m_j > m_{j-1}$ such that $\sum_{k=0}^{n_{j-1}} |a(m_j,k)| < \frac{1}{8}$ (by condition (i)). Choose $n_j > n_{j-1}+1$ such that $\sum_{k=n_j}^\infty |a(m_j,k)| < \frac{1}{8}$ (by condition (iii)).

Define the sequence $\{z_k\}_0^\infty$ by

$$z_k = \begin{cases} 1 & \text{if } n_{2j} \leqslant k \leqslant n_{2j+1} \quad (j = 0, \ldots) \\ \\ 0 & \text{otherwise.} \end{cases}$$

So, the sequence $\{z_k\}_0^\infty$ is bounded (by 1). We now show that the A-transform, $\{\sigma_n\}_0^\infty$, of the sequence $\{z_k\}_0^\infty$ does not satisfy the Cauchy criterion for convergence and, hence, is not convergent.

Let $K$ be any fixed positive integer. Choose $j$ such that $m_{2j} > K$ (hence $m_{2j+1} > K$). Consider

$$|\sigma(m_{2j+1}) - \sigma(m_{2j})| = \left| \sum_{k=0}^\infty a(m_{2j+1},k) z_k - \sum_{k=0}^\infty a(m_{2j},k) z_k \right|$$

$$\geqslant \left[ \left| \sum_{k=0}^\infty a(m_{2j+1},k) z_k \right| \right] - \left[ \sum_{k=0}^\infty |a(m_{2j},k)| \, |z_k| \right]$$

$$= \left[ \left| \sum_{k=0}^{n_{2j}} a(m_{2j+1},k) z_k + \sum_{k=n_{2j}+1}^{n_{2j+1}-1} a(m_{2j+1},k) z_k \right. \right.$$

$$\left. + \sum_{k=n_{2j+1}}^{\infty} a(m_{2j+1},k) z_k \left| \right. \right] - \left[ \sum_{k=0}^{n_{2j}-1} |a(m_{2j},k)| |z_k| \right.$$

$$\left. + \sum_{k=n_{2j-1}+1}^{n_{2j}-1} |a(m_{2j},k)| |z_k| + \sum_{k=n_{2j}}^{\infty} |a(m_{2j},k)| |z_k| \right]$$

$$\geqslant \left| \sum_{k=n_{2j}+1}^{n_{2j+1}-1} a(m_{2j+1},k) \right| - \sum_{k=0}^{n_{2j}} |a(m_{2j+1},k)|$$

$$- \sum_{k=n_{2j+1}}^{\infty} |a(m_{2j+1},k)| - \sum_{k=0}^{n_{2j}-1} |a(m_{2j},k)| - \sum_{k=n_{2j}}^{\infty} |a(m_{2j},k)|$$

$$\geqslant \left| \sum_{k=0}^{\infty} a(m_{2j+1},k) - \sum_{k=0}^{n_{2j}} a(m_{2j+1},k) - \sum_{k=n_{2j+1}}^{\infty} a(m_{2j+1},k) \right|$$

$$- \frac{1}{8} - \frac{1}{8} - \frac{1}{8} - \frac{1}{8}$$

$$\geqslant \left| \sum_{k=0}^{\infty} a(m_{2j+1},k) \right| - \sum_{k=0}^{n_{2j}} |a(m_{2j+1},k)|$$

$$- \sum_{k=n_{2j+1}}^{\infty} |a(m_{2j+1},k)| - \frac{1}{2}$$

$$\geqslant \frac{7}{8} - \frac{1}{8} - \frac{1}{8} - \frac{1}{2} = \frac{1}{8}.$$

Therefore, $\{\sigma(m_k)\}_0^\infty$ is divergent and, hence, the $A$-transform, $\{\sigma_n\}_0^\infty$, diverges. △

We now examine four theorems which demonstrate the existence of regular transformations which satisfy certain given conditions on the behaviour of their transforms.

2.6 THEOREM. *If $\{z_k\}_0^\infty$ is a bounded sequence then there exists a regular matrix $A$ such that the $A$-transform of $\{z_k\}_0^\infty$ converges.*

*Proof.* Since $\{z_k\}_0^\infty$ is bounded there exists a subsequence, $\{z(n_k)\}_0^\infty$, which converges. We may assume, without loss of generality, that $n_0 < n_1 < \ldots < n_k < \ldots$ .
Define the matrix $A = (a_{j,k})$ by

$$a_{j,k} = \begin{cases} 1 & \text{if } k = n_j \ (j=0, 1, \ldots) \\ \\ 0 & \text{otherwise.} \end{cases}$$

Since each row of the matrix has exactly one 1 and the rest zeros and each column has at most one 1 and the rest zeros, we have that conditions (i)-(iii) of Theorem 2.3 are satisfied, i.e. $A$ is regular.

Now, $\sum\limits_{k=0}^{\infty} a_{j,k} z_k = z_{n_j}$ so the $A$-transform of $\{z_k\}_0^{\infty}$ is the subsequence $\{z(n_k)\}_0^{\infty}$ and, therefore, the $A$-transform of $\{z_k\}_0^{\infty}$ converges. $\triangle$

2.7 THEOREM. If $\{z_k\}_0^{\infty}$ and $\{w_k\}_0^{\infty}$ are both bounded sequences and $\{z_k\}_0^{\infty}$ diverges then there exists a regular matrix $A$ such that the $A$-transform of the sequence $\{z_k\}_0^{\infty}$ is the sequence $\{w_k\}_0^{\infty}$.

Proof. Suppose that $\{z_k\}_0^{\infty}$ is a bounded divergent sequence. There exist two distinct points, $\alpha$ and $\beta$, and positive integers $m_0 < n_0 < m_1 < n_1 < \ldots < m_k < n_k < \ldots$ such that $\{z(m_k)\}_0^{\infty}$ converges to $\alpha$, $\{z(n_k)\}_0^{\infty}$ converges to $\beta$, and $z(m_k) \neq z(n_k)$ for all $k = 0, 1, \ldots$. Define the matrix $A = (a_{j,p})$ by

$$
a_{j,p} = \begin{cases} \dfrac{z(n_j) - w_j}{z(n_j) - z(m_j)} & \text{if } p = m_j \\[3mm] \dfrac{w_j - z(m_j)}{z(n_j) - z(m_j)} & \text{if } p = n_j \\[3mm] 0 & \text{otherwise.} \end{cases}
$$

Since each column has at most one non-zero element we have that $\lim_{j\to\infty} a_{j,p} = 0$ for each fixed $p = 0, 1, \ldots$ (thus, condition (i) of Theorem 2.3 is satisfied). Next,

$$
\sum_{p=0}^{\infty} a_{j,p} = \frac{z(n_j) - w_j}{z(n_j) - z(m_j)} + \frac{w_j - z(m_j)}{z(n_j) - z(m_j)} = 1 \quad \text{for each } j = 0,1,\ldots
$$

hence $\lim_{j\to\infty} \sum\limits_{p=0}^{\infty} a_{j,p} = 1$ (which satisfies condition (ii) of Theorem 2.3). Finally, suppose that $|z_k| \leqslant M < \infty$ and $|w_k| \leqslant N < \infty$ for each $k = 0, 1, \ldots$ and for some $M > 0$ and $N > 0$. Thus we have,

$$
\left| \frac{z(n_j) - w_j}{z(n_j) - z(m_j)} \right| \leqslant \frac{|z(n_j)| + |w(j)|}{|z(n_j) - z(m_j)|} \leqslant \frac{M + N}{K}
$$

where $K = \inf_j \{|z(n_j) - z(m_j)|\} > 0$ (since $z(n_j) \neq z(m_j)$, $\{z(m_j)\}_0^{\infty} \to \alpha$, $\{z(n_j)\}_0^{\infty} \to \beta$ and $\alpha \neq \beta$).

Therefore, $\sum\limits_{p=0}^{\infty} |a_{j,p}| \leqslant 2\,[(M+N)/K\,]$ for all $j = 0, 1, \ldots$ and, hence, condition (iii) of Theorem 2.3 holds. Therefore, $A = (a_{j,p})$ is regular. Now, the $A$-transform of $\{z_k\}_0^{\infty}$ is given by

$$\sum_{p=0}^{\infty} a_{j,p} z_p = \frac{z(n_j) - w_j}{z(n_j) - z(m_j)}\,[z(m_j)] + \frac{w_j - z(m_j)}{z(n_j) - z(m_j)}\,[z(n_j)] = w_j$$

which proves the theorem. $\triangle$

2.8 THEOREM. *If $\{z_k\}_0^{\infty}$ is an unbounded sequence and $\{w_k\}_0^{\infty}$ is an arbitrary sequence then there exists a regular matrix $A$ such that the $A$-transform of the sequence $\{z_k\}_0^{\infty}$ is the sequence $\{w_k\}_0^{\infty}$.*

*Proof.* Without loss of generality we may assume that the sequence $\{|z_k|\}_0^{\infty}$ is strictly increasing. So, there exist integers $q_0 < q_1 < \cdots < q_n < \cdots$ such that $q_n > n$ and

$$1 \leqslant \frac{|z(q_n) - w_n| + |w_n - z_n|}{|z(q_n) - z_n|} \leqslant 2 \quad \text{for all } n = 0, 1, \ldots$$

since $|z(q_n) - z_n| \leqslant |z(q_n) - w_n| + |w_n - z_n|$ and, for any fixed $n$, we can choose $q_n > n$ large enough so that

$$\frac{|z(q_n) - w_n|}{|z(q_n) - z_n|} \leqslant \frac{3}{2} \quad \text{and} \quad \frac{|w_n - z_n|}{|z(q_n) - z_n|} \leqslant \frac{1}{2}$$

(because $\{z_k\}_0^{\infty}$ is unbounded).

Define the matrix $A = (a_{j,p})$ by

$$a_{j,p} = \begin{cases} \dfrac{z(q_j) - w_j}{z(q_j) - z_j} & \text{if } p = j \\[2ex] \dfrac{w_j - z_j}{z(q_j) - z_j} & \text{if } p = q_j \\[2ex] 0 & \text{otherwise.} \end{cases}$$

First, $\lim_{j \to \infty} a_{j,p} = 0$ for each fixed $p = 0, 1, \ldots$ since each column of the matrix has at most two non-zero elements.

Next,

$$\lim_{j\to\infty} \sum_{p=0}^{\infty} a_{j,p} = \lim_{j\to\infty} \left( \frac{z(q_j) - w_j}{z(q_j) - z_j} + \frac{w_j - z_j}{z(q_j) - z_j} \right) = 1.$$

Finally,

$$\sum_{p=0}^{\infty} |a_{j,p}| = \left| \frac{z(q_j) - w_j}{z(q_j) - z_j} \right| + \left| \frac{w_j - z_j}{z(q_j) - z_j} \right| \leqslant 2 \quad \text{for all } j = 0, 1, \ldots$$

thus $\sup_j \{ \sum_{p=0}^{\infty} |a_{j,p}| \} \leqslant 2$.    Therefore, $A = (a_{j,p})$ is regular by Theorem 2.3.

Now, examine the $A$-transform of the sequence $\{z_k\}_0^\infty$. We have

$$\sum_{p=0}^{\infty} a_{j,p} z_p = \frac{z(q_j) - w_j}{z(q_j) - z_j} (z_j) + \frac{w_j - z_j}{z(q_j) - z_j} (z(q_j)) = w_j$$

which proves the theorem.    $\triangle$

*Exercise.* Prove that the first line of the proof of Theorem 2.7 is valid.

2.9 THEOREM. *If each of the sequences $\{z_k\}_0^\infty$ and $\{w_k\}_0^\infty$ converges to $z$ and $z_k$ is not identically zero then there exists a regular matrix $A$ such that the $A$-transform of the sequence $\{z_k\}_0^\infty$ is the sequence $\{w_k\}_0^\infty$.*

*Proof.* If $\{z_k\}_0^\infty$ is a constant sequence, say $z_k = z \neq 0$ for all $k$, then take

$$a_{j,p} = \begin{cases} \dfrac{w_j}{z} & \text{if } p = j \\ \\ 0 & \text{otherwise.} \end{cases}$$

If $\{z_k\}_0^\infty$ is not a constant sequence then we choose $A$ such that $\sum_{p=0}^{\infty} a_{j,p} = 1$. In this case, if $\{z_k - z\}_0^\infty$ is transformed into $\{w_k - z\}_0^\infty$ then $\{z_k\}_0^\infty$ will be transformed into $\{w_k\}_0^\infty$. Thus, if $\sum_{p=0}^{\infty} a_{j,p} = 1$ then there is no loss in generality in assuming that $z = 0$. Since $z_k$ is not a constant sequence there exists $k_0$ such that $z(k_0) \neq 0$.

If $\{z_k\}_0^\infty$ has a subsequence which is identically zero then choose an increasing sequence of integers $\{q_j\}_0^\infty$ such that $z(q_j) = 0$ and define

$$a_{j,p} = \begin{cases} \dfrac{w_j}{z(k_0)} & \text{if } p = k_0 \\[3mm] 1 - \dfrac{w_j}{z(k_0)} & \text{if } p = q_j \\[3mm] 0 & \text{otherwise.} \end{cases}$$

Since $\{w_j\}_0^\infty$ converges to zero the matrix $(a_{j,p})$ is regular. Moreover $\sum\limits_{p=0}^\infty a_{j,p} z_p = w_j$ and $\sum\limits_{p=0}^\infty a_{j,p} = 1$.

If $\{z_k\}_0^\infty$ does not have a subsequence which is identically zero then choose an increasing sequence of integers $\{q_j\}_0^\infty$ such that $q_j \neq k_0$ ($j = 0, 1, \ldots$) and such that

$$z(q_j) \neq 0 \quad \text{and} \quad |z(q_{j+1}) - z(q_j)| \geqslant \tfrac{1}{2} |z(q_j)|.$$

Now, define

$$a_{j,p} = \begin{cases} \dfrac{w_j}{z(k_0)} & \text{if } p = k_0 \\[4mm] \dfrac{-z(q_{j+1})}{z(q_j) - z(q_{j+1})}\left(1 - \dfrac{w_j}{z(k_0)}\right) & \text{if } p = q_j \\[4mm] \dfrac{z(q_j)}{z(q_j) - z(q_{j+1})}\left(1 - \dfrac{w_j}{z(k_0)}\right) & \text{if } p = q_{j+1} \\[4mm] 0 & \text{otherwise.} \end{cases}$$

The matrix $(a_{j,p})$ is regular (again, using the fact that $\{w_j\}_0^\infty$ converges to zero),

$$\sum_{p=0}^\infty a_{j,p} z_p = w_j \quad \text{and} \quad \sum_{p=0}^\infty a_{j,p} = 1. \quad \triangle$$

Let us now extend the idea of preserving limits of convergent sequences by means of matrix transformations. To do this, we confine our attention to real transformations, i.e. $a_{n,k}$ is real for all $n = 0, 1, \ldots$ and $k = 0, 1, \ldots$. We examine the case of the real sequence $\{x_n\}_0^\infty$ which diverges to $+\infty$ or to $-\infty$ and when the real transformation $A = (a_{n,k})$ preserves this concept of divergence. In essence, we are examining convergence in the extended real number system. Following this line of thought, we will agree that the

$A$-transform, $\{\sigma_n\}_0^\infty$, of the sequence $\{x_n\}_0^\infty$ exists even though the series $\sum_{k=0}^\infty a_{n,k} x_k$ diverges to $+\infty$ or to $-\infty$ (extension of Definition 2.1) for any, or all, of the values $n = 0, 1, \ldots$ . Of course, if the matrix in question is lower triangular (i.e. $a_{n,k} = 0$ for $k > n$) or is row finite (i.e. has, at most, a finite number of non-zero entries in each row) then the $A$-transform will exist in the sense of Definition 2.1.

2.10 DEFINITION. *The real transform $A$ is totally regular if it is regular and, moreover, given a real sequence $\{x_n\}_0^\infty$ which diverges to $+\infty$ or to $-\infty$ the $A$-transform of $\{x_n\}_0^\infty$ diverges to $+\infty$ or to $-\infty$ respectively.*

2.11 DEFINITION. *The matrix $A = (a_{n,k})$ is positive if there exists $k_0$ such that $a_{n,k} \geqslant 0$ for $k \geqslant k_0$ and all $n = 0, 1, \ldots$ .*

2.12 THEOREM. *If the real transform $A = (a_{n,k})$ is regular and positive then $A$ is totally regular.*

*Proof.* Let $\{x_n\}_0^\infty$ be a real sequence diverging to $+\infty$. Since $A$ is positive there exists $k_0$ such that $a_{n,k} \geqslant 0$ for $k \geqslant k_0$ and all $n = 0, 1, \ldots$ . Let $M > 0$ be given. There exists $k_1 > k_0$ such that $x_k > \frac{8}{7}(M+1)$ for $k \geqslant k_1$. Let $T = \max\{|x_k| : k = 0, \ldots, k_1 - 1\}$. There exists $n_0$ such that if $n \geqslant n_0$ then $a_{n,k} < 1/(T+1)k_1$ for $k = 0, \ldots, k_1 - 1$ (since $A$ is regular). There exists $n_1 \geqslant n_0$ such that if $n \geqslant n_1$ then $\sum_{k=k_1}^\infty a_{n,k} \geqslant \frac{7}{8}$. Now, if $n \geqslant n_1$ then

$$\sigma_n = \sum_{k=0}^\infty a_{n,k} x_k \geqslant \frac{8}{7}(M+1) \sum_{k=k_1}^\infty a_{n,k} - \sum_{k=0}^{k_1-1} a_{n,k} |x_k| \geqslant (M+1) - 1 = M.$$

Thus, $\{\sigma_n\}_0^\infty$ diverges to $+\infty$. The case when $\{x_k\}_0^\infty$ diverges to $-\infty$ is handled similarly. $\triangle$

The condition that $A$ is regular and positive is also necessary in the case when $A$ is a lower triangular matrix as is seen in the following

2.13 THEOREM. *If the real transform, $A = (a_{n,k})$ is lower triangular and totally regular then $A$ is regular and positive.*

*Proof.* The fact that $A$ is regular comes from Definition 2.9. Also, since $A = (a_{n,k})$ is regular, there exists $M > 0$ such that $|a_{n,k}| \leqslant M$ for all $n = 0, 1, \ldots$ and $k = 0, 1, \ldots$ .

Suppose that $A$ is not positive, i.e. given $K$ there exists $k \geqslant K$ such that $a_{n,k} < 0$ for some value of $n$. Since $A$ is lower triangular, there exist sequences of integers $\{n_i\}$ and $\{k(n_i)\}$ such that $a(n_i, k(n_i)) < 0$, $n_i > n_{i-1}$ $(i \geqslant 1)$, and $k(n_i) > k(n_{i-1})$ $(i \geqslant 1)$. Choose a subsequence $\{n_{i,r}\}$ of the sequence $\{n_i\}$ and a sequence $\{s_k\}$ as follows :

Let $n_{i,1} = n_1$ and $s_k = k$ for $k = 0, \ldots, n_{i,1}$.

In general, for $r \geqslant 2$, choose $n_{i,r}$ such that $k(n_{i,r}) > n_{i,r-1}$ (hence $n_{i,r} \geqslant k(n_{i,r}) > n_{i,r-1}$) and

$$\left| \sum_{k=0}^{n_{i,r}-1} a(n_{i,r},k)\, s_k \right| < 1.$$

Choose

$$
s_k = 
\begin{cases}
k & \text{for } n_{i,r-1} + 1 \leqslant k \leqslant n_{i,r},\ k \neq k(n_{i,r}) \\[2ex]
\dfrac{(n_{i,r})^2 (2M)}{|a(n_{i,r}, k(n_{i,r}))|} & \text{for } k = k(n_{i,r}).
\end{cases}
$$

Since $|a_{n,k}| \leqslant M$ for all $n = 0, 1, \ldots$ and $k = 0, 1, \ldots$ we have that the real sequence $\{s_k\}_0^\infty$ diverges to $+\infty$. However,

$$\sum_{k=0}^{\infty} a(n_{i,r+1},k)\, s_k = \sum_{k=0}^{n_{i,r+1}} a(n_{i,r+1},k)\, s_k$$

$$= \sum_{k=0}^{n_{i,r}} a(n_{i,r+1},k)\, s_k + \sum_{k=n_{i,r}+1}^{n_{i,r+1}} a(n_{i,r+1},k)\, s_k$$

$$< 1 + n_{i,r+1} \sum_{k=n_{i,r}+1}^{n_{i,r+1}} |a(n_{i,r+1},k)| - (n_{i,r+1})^2 (2M)$$

$$< 1 + (n_{i,r+1})^2 M - (n_{i,r+1})^2 (2M)$$

$$= 1 - M(n_{i,r+1})^2.$$

Hence, the $A$-transform of $\{s_k\}_0^\infty$ has a subsequence which diverges to $-\infty$. Therefore, $A$ is not totally regular. $\triangle$

Unfortunately, Theorem 2.13 does not hold in general. For example, the matrix $A = (a_{n,k})$ defined by

$$a_{n,k} = \begin{cases} 1 & \text{if } k = n \\ 0 & \text{if } k < n \\ -(\frac{1}{2})^k & \text{if } k = n+1 \\ (\frac{1}{2})^k & \text{if } k > n+1 \end{cases}$$

is totally regular (the $A$-transform $\{\sigma_n\}_0^\infty$ of a sequence $\{x_k\}_0^\infty$ is given by

$$\sigma_n = x_n - (\tfrac{1}{2})^{n+1} x_{n+1} + \sum_{k=2}^\infty (\tfrac{1}{2})^{n+k} x_{n+k}).$$

*Exercise.* Prove the statement: If the real transform $A = (a_{n,k})$ is row finite and totally regular then $A$ is regular and positive.

## Invertibility

The concept of finding an inverse of a matrix is an extremely important one. We shall discover later, in this chapter and in the following chapters, that being able to find an inverse will allow us to formulate and to prove many results.

2.14 DEFINITION. *Let* $A = (a_{n,k})$ *be an arbitrary matrix. The matrix* $B = (b_{n,k})$ *is a left-hand inverse of the matrix* $A$ *if*

$$\sum_{k=0}^\infty b_{n,k} a_{k,j} = \begin{cases} 1 & \text{if } n = j \\ 0, & \text{if } n \neq j, \end{cases}$$

*in which case, $A$ is left invertible. The matrix $B$ is a right-hand inverse of the matrix $A$ if*

$$\sum_{k=0}^\infty a_{n,k} b_{k,j} = \begin{cases} 1 & \text{if } n = j \\ 0 & \text{if } n \neq j, \end{cases}$$

*in which case $A$ is right invertible. If $A$ is both left and right-invertible then $A$ is invertible.*

Unfortunately, as compared to the case with finite $n \times n$ matrices, a matrix $A$ may be left (right) invertible, but not right (left) invertible. Also, a matrix $A$ may have more than one left (right) hand inverse. In the following, we will see that we can say what will happen in the case of lower triangular matrices. In general, the question of invertibility depends greatly upon associativity of multiplication.

2.15 DEFINITION. *Let* $A = (a_{n,k})$ *and* $B = (b_{n,k})$ *be two arbitrary matrices. The product $AB$ is the matrix $C = (c_{n,k})$ where*

$c_{n,k} = \sum_{j=0}^{\infty} a_{n,j} \, b_{j,k}$ provided this sum exists for all $n$, $k = 0, 1, \ldots$.
Otherwise the product does not exist.

2.16  DEFINITION.  Let $A$, $B$, and $C$ be arbitrary matrices.  The product $(AB)C$ is associative if $(AB)C = A(BC)$.

2.17  THEOREM.  (i) If $A$ and $B$ are each lower triangular matrices and $C$ is an arbitrary matrix, then $(AB)C = A(BC)$.

(ii) If $A$ and $B$ are each lower triangular matrices and $C$ is a column matrix (i.e. $C$ has only one column) then $(AB)C = A(BC)$.

Proof.  The proof of (i) is a direct calculation and the proof of (ii) follows by letting

$$c_{n,k} = \begin{cases} c_n & \text{for } k = 0 \\ 0 & \text{for } k \neq 0 \end{cases}$$

in part (i) and comparing $(c_{n,k})$ to a column matrix.   △

2.18  THEOREM.  A lower triangular matrix $A = (a_{n,k})$ is invertible if and only if $a_{n,n} \neq 0$ for all $n = 0, 1, \ldots$.   Moreover, if $A$ is invertible then it has a unique right hand inverse which is also a left hand inverse.

Proof.  First, assume that $A$ is invertible.  By solving the system of equations

$$\sum_{j=0}^{n} a_{n,j} \, b_{j,k} = \begin{cases} 1 & \text{if } n = k \\ 0 & \text{if } n \neq k \end{cases} \tag{2.4}$$

we see that any right-hand inverse, $B = (b_{n,k})$, is lower triangular and $a_{n,n} b_{n,n} = 1$ for each fixed $n$, i.e. $a_{n,n} \neq 0$ for each fixed $n$. Now, since $a_{n,n} \neq 0$, we can construct a lower triangular left-hand inverse, $C = (c_{n,k})$ (where $c_{n,k} = 0$ for $k > n$) by solving the system of equations

$$\sum_{j=k}^{n} c_{n,j} \, a_{j,k} = \begin{cases} 1 & \text{if } n = k \\ 0 & \text{if } k < n. \end{cases} \tag{2.5}$$

If $B$ and $B'$ are any right-hand inverses then, by Theorem 2.16, $B = (CA)B = C(AB) = C$ and $B' = (CA)B' = C(AB') = C$, i.e. $B = B'$ and $B$ is a left-hand inverse.

Now, assume that $a_{n,n} \neq 0$ for each fixed $n$.   By solving the system of equations (2.1) we obtain a right-hand inverse $B$ and by assuming $c_{n,k} = 0$ for $k > n$ and solving the system of equations (2.5) we obtain a left-hand inverse. Hence, $A$ is invertible. $\triangle$

We note that Theorem 2.17 does not imply that a lower triangular matrix with non-zero diagonal elements has a unique left-hand inverse. For example, let $A = (a_{n,k})$ be defined by

$$a_{n,k} = \begin{cases} 1 & \text{if } k = n-1 \text{ or } k = n \quad (n \geqslant 1) \\ 1 & \text{if } k = n = 0 \\ 0 & \text{otherwise.} \end{cases}$$

The unique right-hand inverse is $B = (b_{n,k})$ where

$$b_{n,k} = \begin{cases} 0 & \text{if } k > n \\ (-1)^{n+k} & \text{if } k \leqslant n. \end{cases}$$

So $B$ is also a left-hand inverse. However, the matrix $C = (c_{n,k})$ where

$$c_{n,k} = \begin{cases} (-1)^{n+k+1}\,\tfrac{1}{2} & \text{if } k \geqslant n+1 \\ (-1)^{n+k}\,\tfrac{1}{2} & \text{if } k \leqslant n \end{cases}$$

is also a left-hand inverse.

We can, however, say

2.19 THEOREM. *An invertible lower triangular matrix has only one lower triangular left-hand inverse.*

*Proof.* Let $A$ be an invertible lower triangular matrix. By Theorem 2.18, there exists a unique right-hand inverse $B$ which is lower triangular and is a left-hand inverse also. Suppose that $C$ is a lower triangular left-hand inverse. We have that

$$C = C(AB) = (CA)B = B$$

by Theorem 2.17. $\triangle$

We conclude this section by giving an example of a matrix $A$ (not lower triangular) which has no right-hand inverse, but it does have an infinity of left-hand inverses. Define $A = (a_{n,k})$ by

$$a_{n,k} = \begin{cases} 1 & \text{if } n = k = 0 \\ 1 & \text{if } k = n+1 \text{ or } k = n-1 \text{ and } n \geqslant 1 \\ 0 & \text{otherwise.} \end{cases}$$

We leave the calculations to the reader.

**Inclusion**

Let $A$ and $B$ be two summability transforms and let $c_A$ and $c_B$ be the collections of sequences which are $A$-summable and $B$-summable respectively. The question of inclusion is one of examining when $c_A$ is a subset of $c_B$ and, moreover, given any sequence in $c_A$ it is $A$-summable and $B$-summable to the same value. Of particular interest is the case when $A$ and $B$ are both regular (in which case any convergent sequence is $A$-summable and $B$-summable to the same value). The question of inclusion, in this case, is restricted to examining the summability of divergent sequences.

2.20 DEFINITION. *Let $A$ and $B$ be transforms.*
(i) *If the sequence $\{z_k\}_0^\infty$ is both $A$-summable and $B$-summable then $A$ and $B$ are comparable for $\{z_k\}_0^\infty$.*
(ii) *If, given any sequence for which $A$ and $B$ are comparable, the $A$ transform converges to the same value as does the $B$ transform then $A$ and $B$ are consistent. Otherwise $A$ and $B$ are inconsistent.*
(iii) *If every sequence which is $A$-summable is also $B$-summable and $A$ and $B$ are consistent then $A$ is included in $B$ (written $A \subset B$).*
(iv) *If $A \subset B$ and $B \subset A$ then $A$ and $B$ are mutually consistent (or equivalent) over the entire space of sequences.*

2.21 THEOREM. *Let $A = (a_{n,k})$ and $B = (b_{n,k})$ be transforms. If*
(i)   *$A$ has a left-hand inverse (say $C = (c_{n,k})$),*
(ii)  *$\displaystyle\sum_{j=0}^\infty \sum_{k=0}^\infty |c_{n,j}\, a_{j,k}\, z_k|$ converges for every sequence $\{z_k\}_0^\infty$ which is $A$-summable,*
(iii) *$\displaystyle\sum_{j=0}^\infty \sum_{k=0}^\infty |b_{n,k}\, c_{k,j}|$ converges and*
(iv)  *the matrix $D = (d_{n,k})$ defined by*

$$d_{n,k} = \sum_{j=0}^\infty b_{n,j}\, c_{j,k}$$

*is regular then $A \subset B$.*

*Proof.* Let $\{z_k\}_0^\infty$ be $A$-summable to the value $y$, i.e. the sequence $\{\sigma_n\}_0^\infty$ converges to $y$ where

$$\sigma_n = \sum_{k=0}^\infty a_{n,k}\, z_k.$$

We have that

$$\sum_{j=0}^\infty c_{n,j}\, \sigma_n = \sum_{j=0}^\infty c_{n,j}\left( \sum_{k=0}^\infty a_{j,k}\, z_k \right)$$

$$= \sum_{k=0}^\infty \left( \sum_{j=0}^\infty c_{n,j}\, a_{j,k} \right) z_k = z_n$$

by conditions (i) and (ii). Thus, the $B$-transform, $\{\tau_n\}_0^\infty$, of the sequence $\{z_k\}_0^\infty$ is given by

$$\tau_n = \sum_{k=0}^\infty b_{n,k} z_k = \sum_{k=0}^\infty b_{n,k} \left( \sum_{j=0}^\infty c_{k,j} \sigma_j \right).$$

By condition (iii) and the fact that $\{\sigma_j\}_0^\infty$ is bounded we have that

$$\tau_n = \sum_{j=0}^\infty \left( \sum_{k=0}^\infty b_{n,k} c_{k,j} \right) \sigma_j = \sum_{j=0}^\infty d_{n,j} \sigma_j,$$

i.e. $\{\tau_n\}_0^\infty$ is merely the $D$-transform of the convergent sequence $\{\sigma_j\}_0^\infty$. By condition (iv) it follows that $\{\tau_n\}_0^\infty$ converges to $y$ also, i.e. $A \subset B$. $\triangle$

2.22 THEOREM. *If $A$ is regular, lower triangular, and $m > n$ (both positive integers) then $A^n \subset A^m$.*

*Proof.* It is sufficient to prove that $A^h \subset A^{h+1}$ (this will give us the fact that $A^n \subset A^{n+1} \subset \ldots \subset A^m$). Let $\{z_k\}_0^\infty$ be $A^h$-summable to $y$ and let $\{u_k\}_0^\infty$ be the $A^h$-transform of $\{z_k\}_0^\infty$. As in the proof of Theorem 2.4 we have that the $A^{h+1}$-transform of $\{z_k\}_0^\infty$ is merely the $A$-transform of $\{u_k\}_0^\infty$ and, since $A$ is regular and $\{u_k\}_0^\infty$ converges to $y$, we see that the $A^{h+1}$-transform of $\{z_k\}_0^\infty$ converges to $y$. Therefore $A^h \subset A^{h+1}$. $\triangle$

Theorem 2.22 does not hold in general. Wilansky and Zeller (*J. Lond. Math. Soc.* **32** (1957), 397–408 (Theorem 7)) gave an example of a regular matrix $A$ which sums a divergent sequence but $A^2$ is the identity matrix (thus it is false that $A^2 \supset A$).

2.23 DEFINITION. *The transforms $A = (a_{n,k})$ and $B = (b_{n,k})$ are absolutely equivalent over the arbitrary collection of sequences $S$ if, given $\{z_k\}_0^\infty \in S$ and*

$$\sigma_n = \sum_{k=0}^\infty a_{n,k} z_k \quad \text{and} \quad \tau_n = \sum_{k=0}^\infty b_{n,k} z_k,$$

*then the sequence $\{w_n\}_0^\infty$ converges to 0 where $w_n = \sigma_n - \tau_n$.*

Clearly, absolute equivalence over the entire space of sequences implies mutual consistency over this space.

*Exercise.* Prove that mutual consistency over the entire space of sequences does not imply absolute equivalence over this space.

## Translativity

We conclude this chapter with a discussion of an interesting phenomenom which occurs in the study of summability transforms. Let $z = \{z_k\}_0^\infty$ be any sequence and define the sequence $w = \{w_k\}_0^\infty$

by $w_k = z_{k+1}$ ($k = 0, 1, \ldots$ ). The convergence of the sequence $z$ implies the convergence of the sequence $w$ (to the same value) and conversely. One might expect that, given a transform $A$, if $z$ is $A$-summable to $s$ then $w$ is $A$-summable to $s$ and conversely. This, however, is not necessarily the case (and is a topic for much research).

2.24   DEFINITION. *For any sequence $z$ if $z$ is $A$-summable to $s$ implies that $w$ is $A$-summable to $s$ then $A$ is right translative. If $w$ is $A$-summable to $s$ implies that $z$ is $A$-summable to $s$ then $A$ is left translative. If $A$ is both right and left translative then $A$ is translative. If the possibilities include $s = \pm \infty$ then $A$ is right totally translative, left totally translative, and totally translative respectively.*

A method of attacking the problem of translativity, which is similar to that method presented (in Theorem 2.21) for examining inclusion, is now given. As is readily seen this method (and the one used in Theorem 2.21) is quite restrictive due to the many limitations we place upon the transforms in question.

2.25   THEOREM. *Let $A = (a_{n,k})$ be a transform. If*
(i)   *$A$ has a left-hand inverse (say $B = (b_{n,k})$),*

(ii)   $\displaystyle\sum_{j=0}^{\infty} \sum_{k=0}^{\infty} |b_{n,j} \, a_{j,k} \, z_k|$ *converges for every sequence $\{z_k\}_0^{\infty}$ which is $A$-summable,*

(iii)   $\displaystyle\sum_{k=1}^{\infty} \sum_{j=0}^{\infty} |a_{n,k-1} \, b_{k,j}|$ *converges, and*

(iv)   *the matrix $D = (d_{n,k})$ defined by*

$$d_{n,k} = \sum_{j=1}^{\infty} a_{n,j-1} \, b_{j,k}$$

*is regular (totally regular)*
*then $A$ is right translative (right totally translative).*

*Proof.* Let $\{z_k\}_0^{\infty}$ be $A$-summable, let $\{\sigma_n\}_0^{\infty}$ be the $A$ transform of $z = \{z_k\}_0^{\infty}$, i.e.

$$\sigma_n = \sum_{k=0}^{\infty} a_{n,k} \, z_k,$$

and let $\{\tau_n\}_0^{\infty}$ be the $A$ transform of the sequence $w = \{w_k\}_0^{\infty}$, i.e.

$$\tau_n = \sum_{k=0}^{\infty} a_{n,k} \, w_k = \sum_{k=1}^{\infty} a_{n,k-1} \, z_k.$$

By conditions (i) and (ii) we have that

$$z_n = \sum_{k=0}^{\infty} b_{n,k} \, \sigma_n.$$

Thus, by condition (iii),

$$\tau_n = \sum_{k=1}^{\infty} a_{n,k-1} \left( \sum_{j=0}^{\infty} b_{k,j} \, \sigma_j \right)$$

$$= \sum_{j=0}^{\infty} \left( \sum_{k=1}^{\infty} a_{n,k-1} \, b_{k,j} \right) \sigma_j = \sum_{j=0}^{\infty} d_{n,j} \, \sigma_j .$$

Hence, if $D$ is regular then $A$ is right translative and if $D$ is totally regular then $A$ is right totally translative. $\triangle$

*Exercise.* Find a matrix which is not right translative.

*Exercise.* Discuss left translativity and left total translativity.

If we restrict ourselves to consideration of bounded sequences only in the study of inclusion and translativity then condition (ii) in Theorem 2.21 becomes

$$\sum_{j=0}^{\infty} \sum_{k=0}^{\infty} \left| c_{n,j} \, a_{j,k} \right| \text{ converges}$$

and condition (ii) in Theorem 2.25 becomes

$$\sum_{j=0}^{\infty} \sum_{k=0}^{\infty} \left| b_{n,j} \, a_{j,k} \right| \text{ converges.}$$

CHAPTER 3


## Well-known Methods of Summability


In this chapter various well-known methods of summability are surveyed along with an examination of some properties of these methods. Also, we will make a few historical remarks as we introduce the various methods.

### Nörlund and Nörlund-type translations

A Russian mathematician, Voronoi, first defined a Nörlund mean (see Definition 3.1) in the *Proceedings of the eleventh congress of Russian naturalists and scientists*, St. Petersburg (pp. 60-1) in 1902. This went unnoticed until much later. Independently of Voronoi, N. E. Nörlund (sometimes written Nørlund) gave the definition in *Acta Univ. lund* (2), volume 16, no. 3 in 1920.

3.1 DEFINITION. *Let* $\{q_n\}_0^\infty$ *be a sequence of non-negative real numbers with* $q_0 > 0$. *Define* $Q_n = \sum\limits_{k=0}^{n} q_k$.

(i) *The Nörlund transformation* $(N, q_n) = (a_{n,k})$ *is defined by*

$$a_{n,k} = \begin{cases} q_{n-k}/Q_n & \text{if } k \leqslant n \\ 0 & \text{if } k > n. \end{cases}$$

(ii) *The Nörlund type transformation* $(R, q_n) = (r_{n,k})$ *is defined by*

$$r_{n,k} = \begin{cases} q_k/Q_n & \text{if } k \leqslant n \\ 0 & \text{if } k > n. \end{cases}$$

If $q_n = 1$ for each $n = 0, 1, \ldots$ then we have the $(C,1)$ transformation (in both the Nörlund and Nörlund-type transformations) which we examined in Chapter 1.

3.2 THEOREM. *The* $(N, q_n)$ *transformation is regular if and only if* $\lim_{n \to \infty} \dfrac{q_n}{Q_n} = 0.$

*Proof.* The $(N, q_n)$ transformation is regular if and only if conditions (i)–(iii) of Theorem 2.3 hold. Conditions (ii) and (iii) hold since $|a_{n,k}| = a_{n,k}$ for all $n = 0, 1, \ldots$ and $k = 0, 1, \ldots$ and

$$\sum_{k=0}^{\infty} a_{n,k} = \sum_{k=0}^{n} a_{n,k} = \sum_{k=0}^{n} \frac{q_{n-k}}{Q_n} = \frac{1}{Q_n} \sum_{k=0}^{n} q_{n-k} = 1.$$

Now, let $k$ be fixed. We have (for $k = 0$) that $\lim_{n \to \infty} a_{n,0} = \lim_{n \to \infty} q_n/Q_n$ thus, if $(N, q_n)$ is regular then $\lim_{n \to \infty} q_n/Q_n = 0$. Now, suppose that $\lim_{n \to \infty} q_n/Q_n = 0$. We have, for $n \geqslant k$ and $k$ fixed,

$$0 \leqslant a_{n,k} = \frac{q_{n-k}}{Q_n} \leqslant \frac{q_{n-k}}{Q_{n-k}}$$

(since the sequence $\{Q_j\}_0^{\infty}$ is increasing). Therefore,

$$0 \leqslant \lim_{n \to \infty} a_{n,k} \leqslant \lim_{n \to \infty} \frac{q_{n-k}}{Q_{n-k}} = 0,$$

i.e. $\lim_{n \to \infty} a_{n,k} = 0$ and, thus, condition (i) of Theorem 2.3 holds, which implies that $(N, q_n)$ is regular. $\triangle$

*Exercise.* (i) Prove that the $(R, q_n)$ transformation is regular if and only if $\lim_{n \to \infty} Q_n = \infty$.
(ii) Determine a lower triangular matrix which is both a left-hand and a right-hand inverse of

(a) the $(N, q_n)$ transform

(b) the $(R, q_n)$ transform.

### Hölder means and Cesàro means

In examining the Hölder and Cesàro transformations (frequently referred to as the Hölder means and the Cesàro means) we will see that both are generalizations of the $(C, 1)$ transform (which was discussed in Chapter 1).

O. Hölder first defined the $n$th Hölder mean (see Definition 3.3) in 'Grenzwerte von Reihen an der ·Konvergenzgrenze', *Mathematische Annalen*, volume 20, pages 535-49 in 1882. For obvious reasons the Hölder mean is known as a type of arithmetic mean.

3.3  DEFINITION.  *The first Hölder mean, $(H,1) = (h_{n,\ k}^{\ 1})$, is defined by*

$$h_{n,k}^1 = \begin{cases} \dfrac{1}{n+1} & \text{if } n \geqslant k \\[2mm] 0 & \text{if } n < k. \end{cases}$$

*The mth Hölder mean, $(H,m) = (h_{n,k}^m)$, for m a positive integer, is defined by $(h_{n,\ k}^{\ m}) = (h_{n,k}^1)\ (h_{n,k}^{m-1})$ (where the product of the two matrices denotes usual matrix multiplication).*

The first Hölder mean is, of course, the $(C,1)$ transform.

3.4  THEOREM.  *Let m be a positive integer.   The mth Hölder mean, $(H,m)$, is regular.*

The proof of this theorem follows directly from Theorem 2.3.

3.5  THEOREM.  *If $m > n$ (both positive integers) then $(H,n) \subset (H,m)$.*

The proof of this theorem follows directly from Theorem 2.22.

We noted in Chapter 1 that E. Cesàro first introduced the $(C,k)$ methods, $k$ a positive integer, (see Definition 3.6) in 'Sur la multiplication des séries', *Bulletin des Sciences Mathématiques* (2), volume 14, pages 114-20 in 1890 (the $(C,k)$ methods are further examples of arithmetic means). The more general $(C,\alpha)$ methods (where $\alpha > -1$) were introduced by K. Knopp in *Sitzungsberichte d. Berliner Math. Ges.*, volume 7, pages 1-12 in 1907 and, independently, by S. Chapman in an article in the *Proceedings of the London Mathematical Society* (2), volume 9, pages 360-409 in 1911.

3.6  THEOREM.  *Let $\{z_k\}_0^\infty$ be a sequence of complex numbers. Let $\alpha$ be any real number excluding the negative integers. Define the sequences $\{A_n^\alpha\}_0^\infty$ and $\{S_n^\alpha\}_0^\infty$ by*

$$\sum_{k=0}^\infty A_k^\alpha x^k = (1-x)^{-\alpha-1} \quad \text{and} \quad \sum_{k=0}^\infty S_k^\alpha x^k = (1-x)^{-\alpha} \sum_{k=0}^\infty z_k x^k.$$

*The sequence $\{z_k\}_0^\infty$ is $(C,\alpha)$-summable to $y$ provided $\lim_{k\to\infty} S_k^\alpha / A_k^\alpha = y$.   The $(C,\alpha)$-transform is called the Cesàro mean of order $\alpha$.*

Before we examine the regularity of $(C,\alpha)$, in general, let us comment on the case when $\alpha$ is a positive integer, say $\alpha = m$. We have for $|x| < 1$ (by the binomial theorem) that

$$(1-x)^{-m-1} = \sum_{k=0}^\infty \binom{k+m}{m} x^k$$

where the binomial coefficient $\binom{k}{m}$ has its usual meaning, i.e.

$\binom{k}{m} = \dfrac{k!}{m!\,(k-m)!}$ for $k \geqslant m$). Therefore, $A_k^m = \binom{k+m}{m}$. Also, we have

$$\sum_{k=0}^{\infty} S_k^m x^k = (1-x)^{-m} \sum_{k=0}^{\infty} z_k x^k = \sum_{j=0}^{\infty} \binom{j+m-1}{m-1} x^j \sum_{k=0}^{\infty} z_k x^k$$

$$= \sum_{k=0}^{\infty} c_k x^k \text{ where } c_k = \sum_{t=0}^{k} \binom{k-t+m-1}{m-1} z_t$$

(the Cauchy coefficients of the Cauchy product). Hence $S_k^m = \sum_{t=0}^{k} \binom{k-t+m-1}{m-1} z_t$. So, if we define the matrix $(c_{n,\,k}^{\,m})$ by

$$c_{n,\,k}^{\,m} = \begin{cases} \dbinom{n-k+m-1}{m-1} \Big/ \dbinom{n+m}{m} & \text{if } k \leqslant n \\[2mm] 0 & \text{if } n < k, \end{cases}$$

we have that the $(C,m)$-transform of the sequence $\{z_k\}_0^{\infty}$ is just the $(c_{n,\,k}^{\,m})$ matrix transformation of $\{z_k\}_0^{\infty}$. In particular, if $m = 1$ we have that

$$c_{n,\,k}^{\,1} = \binom{n-k}{0} \Big/ \binom{n+1}{1} = \frac{1}{n+1} \text{ if } k \leqslant n \text{ and } c_{n,\,k}^{\,1} = 0 \text{ if } k > n;$$

i.e. for $m = 1$ we obtain the $(C,1)$-transform (see Definition 1.9) which we discussed in Chapter 1 (of course, this is desirable since we are calling these new transformations by the name Cesàro transformations).

In general (for $\alpha$ any real number excluding the negative integers) we have

$$f(x) = (1-x)^{-\alpha-1} = \sum_{k=0}^{\infty} \frac{f^{(k)}(0)}{k!} x^k = 1 + \sum_{k=1}^{\infty} \frac{(\alpha+1)\ldots(\alpha+k)}{k!} x^k$$

and hence $A_0^{\alpha} = 1$ and $A_1^{\alpha} = \dfrac{(\alpha+1)\ldots(\alpha+k)}{k!}$ for $k \geqslant 1$. Also, we have

$$\sum_{k=0}^{\infty} S_k^{\alpha} x^k = (1-x)^{-\alpha} \sum_{k=0}^{\infty} z_k x^k$$

$$= \left(1 + \sum_{k=1}^{\infty} \frac{\alpha \ldots (\alpha+k-1)}{k!} x^k\right) \left(\sum_{j=0}^{\infty} z_j x^j\right)$$

$$= \sum_{k=0}^{\infty} c_k^{\alpha} x^k$$

where $\qquad c_k^{\alpha} = \sum_{j=0}^{k-1} \dfrac{\alpha\,(\alpha+1)\ldots(\alpha+k-j-1)}{(k-j)!} z_j + z_k.$ $\qquad\qquad$ (3.1)

Therefore, if we define matrix $(c_{k,j}^{\alpha})$ by

$$
c_{k,j}^{\alpha} = \begin{cases}
1 & \text{if } k = j = 0 \\[2mm]
\dfrac{1}{\left[\dfrac{(\alpha+1)\,\cdots\,(\alpha+k)}{k!}\right]} & \text{if } k = j \neq 0 \\[4mm]
\dfrac{\left[\dfrac{\alpha\,\cdots\,(\alpha+k-j-1)}{(k-j)!}\right]}{\left[\dfrac{(\alpha+1)\,\cdots\,(\alpha+k)}{k!}\right]} & \text{if } k > j \\[4mm]
0 & \text{otherwise}
\end{cases}
$$

then the $(C,\alpha)$-transform of the sequence $\{z_k\}_0^{\infty}$ is just the $(c_{k,j}^{\alpha})$ matrix transform of $\{z_k\}_0^{\infty}$.

3.7 **THEOREM.** *If $\alpha > 0$ then $(C,\alpha) = (c_{k,j}^{\alpha})$ is regular.*

*Proof.* First, if $\alpha > 0$ then $c_{k,j}^{\alpha} \geqslant 0$ for all $k = 0, 1, \ldots$ and $j = 0, 1, \ldots$ and, thus $\sum_{j=0}^{\infty} c_{k,j}^{\alpha} = \sum_{j=0}^{\infty} |c_{k,j}^{\alpha}|$. So if we prove that $\lim_{k\to\infty} \sum_{j=0}^{\infty} c_{k,j}^{\alpha} = 1$ (condition (ii) of Theorem 2.3) then $\sup_k \{\sum_{j=0}^{\infty} |c_{k,j}^{\alpha}|\}$ is finite (condition (iii) of Theorem 2.3). Examine $\sum_{j=0}^{\infty} c_{k,j}^{\alpha} = \sum_{j=0}^{k} c_{k,j}^{\alpha}$. If $k = 0$ then $\sum_{j=0}^{0} c_{0,j}^{\alpha} = 1$. If $k \geqslant 1$ we have

$$
\sum_{j=0}^{k} c_{k,j}^{\alpha} = \frac{1}{A_k^{\alpha}}\left[1 + \sum_{j=0}^{k-1} \frac{\alpha\,\cdots\,(\alpha+k-j-1)}{(k-j)!}\right].
$$

Consider $\quad 1 + \sum_{j=0}^{k-1} \dfrac{\alpha\,\cdots\,(\alpha+k-j-1)}{(k=j)!} = c_k^{\alpha}$

in Eq. (3.1) (preceding the statement of this theorem) where the sequence $\{z_n\}_0^{\infty}$ is given by $z_n = 1$ for all $n = 0, 1, \ldots$ . We have

$$
\sum_{k=0}^{\infty} c_k^{\alpha}x^k = \sum_{k=0}^{\infty} S_k^{\alpha}x^k = (1-x)^{-\alpha} \sum_{k=0}^{\infty} z_k x^k = (1-x)^{-\alpha} \sum_{k=0}^{\infty} (1)x^k
$$

$$
= (1-x)^{-\alpha} \sum_{k=0}^{\infty} x^k = (1-x)^{-\alpha}(1-x)^{-1}
$$

$$
= (1-x)^{-\alpha-1} = \sum_{k=0}^{\infty} A_k^{\alpha}x^k,
$$

i.e. $c_k^{\alpha} = A_k^{\alpha}$ and, hence,

$$\sum_{j=0}^{\infty} c_{k,j}^{\alpha} = \sum_{j=0}^{k} c_{k,j}^{\alpha} = \frac{1}{A_k^{\alpha}} \, c_k^{\alpha} \equiv \frac{1}{A_k^{\alpha}} \, A_k^{\alpha} = 1.$$

Therefore, $\lim_{k\to\infty} \sum_{j=0}^{\infty} c_{k,j}^{\alpha} = 1$ and conditions (ii) and (iii) of Theorem 2.3 hold.

Now, it remains to show that $\lim_{k\to\infty} c_{k,j}^{\alpha} = 0$ for each fixed $j = 0, 1, \ldots$ . Let $j$ be fixed and take $k \geqslant j$. We have

$$\lim_{k\to\infty} c_{k,j}^{\alpha} = \lim_{k\to\infty} \frac{\alpha \ldots (\alpha+k-j-1)}{(k-j)!} \, \frac{k!}{(\alpha+1)\ldots(\alpha+k)}$$

$$= \lim_{k\to\infty} \frac{\alpha k \, (k-1) \ldots (k-j+1)}{(\alpha+k-j)\ldots(\alpha+k)}$$

$$= \lim_{k\to\infty} \frac{\alpha k^j}{k^{j+1}} = \lim_{k\to\infty} \frac{\alpha}{k} = 0.$$

Therefore, condition (i) of Theorem 2.3 holds and, hence, $(C,\alpha)$ is regular for $\alpha > 0$. $\triangle$

The condition that $\alpha$ is positive relieved us of examining the boundedness of $\sup_k \{ \sum_{j=0}^{\infty} |c_{k,j}^{\alpha}| \}$. The case when $\alpha = 0$ gives us our original definition of convergence since, in Definition 3.6, $A_k^0 = 1$ for each $k = 0, 1, \ldots$ and $\sum_{k=0}^{\infty} S_k^0 x^k = \sum_{k=0}^{\infty} z_k x^k$, i.e. $S_k^0 = z_k$ for each $k = 0, 1, \ldots$ . Therefore, $\lim_{k\to\infty} S_k^0/A_k^0 = \lim_{k\to\infty} z_k$ which exists if and only if the sequence $\{z_k\}_0^{\infty}$ converges.

We now want to investigate the problem of inclusion of the Cesàro means. To do this we need the following three preparatory statements.

3.8 LEMMA. *If $\alpha$ and $\beta$ are real numbers then*

$$\frac{(\alpha+\beta)\ldots(\alpha+\beta+k-t-1)}{(k-t)!} = \frac{\beta\ldots(\beta-1+k-t)}{(k-t)!}$$

$$+ \sum_{j=t+1}^{k-1} \frac{\alpha\ldots(\alpha-1+j-t)}{(j-t)!} \, \frac{\beta\ldots(\beta-1+k-j)}{(k-j)!}$$

$$+ \frac{\alpha\ldots(\alpha-1+k-t)}{(k-t)!}$$

for $k > t + 1$.

*Proof.* We have $(1-x)^{-(\alpha+\beta)} = 1 + \sum_{k=t+1}^{\infty} \frac{(\alpha+\beta)\ldots(\alpha+\beta+k-t-1)}{(k-t)!} x^{k-t}$.

Also,

$$(1-x)^{-(\alpha+\beta)} = (1-x)^{-\alpha} (1-x)^{-\beta}$$

$$= (1 + \sum_{m=1}^{\infty} \frac{\alpha \ldots (\alpha-1+m)}{m!} x^m) (1 + \sum_{n=1}^{\infty} \frac{\beta \ldots (\beta-1+n)}{n!} x^n)$$

$$= 1 + \sum_{m=1}^{\infty} c_m x^m$$

where, for $m \geqslant 2$ we have

$$c_m = \frac{\beta \ldots (\beta-1+m)}{m!} + \sum_{n=1}^{m-1} \frac{\alpha \ldots (\alpha-1+n)}{n!} \frac{\beta \ldots (\beta-1+m-n)}{(m-n)!}$$

$$+ \frac{\alpha \ldots (\alpha-1+m)}{m!}$$

(note that $c_1 = \alpha + \beta$ also). The result follows by letting $m = k - t$ and $n = j - t$. $\triangle$

3.9  COROLLARY.  *If $k > t + 1$ then*

$$0 = \frac{(-\alpha) \ldots (-\alpha-1+k-t)}{(k-t)!} + \sum_{j=t+1}^{\infty} \frac{\alpha \ldots (\alpha-1+j-t)}{(j-t)!} \frac{(-\alpha) \ldots (-\alpha-1+k-j)}{(k-j)!}$$

$$+ \frac{\alpha \ldots (\alpha-1+k-t)}{(k-t)!} .$$

*Proof.*  Let $\beta = -\alpha$ in Lemma 3.8.  $\triangle$

3.10  LEMMA.  *Define the matrix $(d_{k,j}^{\alpha})$ by $d_{k,j}^{\alpha} = A_k^{\alpha} c_{k,j}^{\alpha}$, i.e.*

$$d_{k,j}^{\alpha} = \begin{cases} 1 & \text{if } k = j \\ \dfrac{\alpha \ldots (\alpha+k-j-1)}{(k-j)!} & \text{if } k > j \\ 0 & \text{otherwise.} \end{cases}$$

*If $(e_{k,t}^{\alpha}) = (d_{k,j}^{-\alpha}) (d_{j,t}^{\alpha})$ then $e_{k,t}^{\alpha} = \begin{cases} 1 & \text{if } k = t \\ 0 & \text{if } k \neq t. \end{cases}$*

*Proof.*  If $k < t$ then, clearly, $e_{k,t}^{\alpha} = 0$ and if $k = t$ then $e_{k,t}^{\alpha} = 1$. Now, for $k = t+1$ we have $e_{k,t}^{\alpha} = \alpha + (-\alpha) = 0$. Let $k > t+1$. Thus

$$e_{k,\,t}^{\,\alpha} = \frac{(-\alpha)\,\ldots\,(-\alpha+k-t-1)}{(k-t)!} + \sum_{j=t+1}^{k-1} \frac{(-\alpha)\,\ldots\,(-\alpha+k-j-1)}{(k-j)!} \frac{\alpha\,\ldots\,(\alpha+j-t-1)}{(j-t)!}$$

$$+ \frac{\alpha\,\ldots\,(\alpha+k-t-1)}{(k-t)!}$$

and, by Corollary 3.9, $e_{k,\,t}^{\,\alpha} = 0$.  $\triangle$

*Exercise.* (i) Prove that $1 - x \leqslant e^{-x}$ for $x \geqslant 0$.

(ii) If $0 \leqslant b_j < 1$ for all $j = 1, 2, \ldots$ then

(a) $(1-b_1)\,\ldots\,(1-b_k) \leqslant \exp(-\sum_{j=1}^{k} b_j)$ and

(b) $\lim_{k\to\infty} (1-b_1)\,\ldots\,(1-b_k) = 0$ provided $\sum_{k=1}^{\infty} b_k$ diverges.

The major result concerning inclusion of Cesàro means is

**3.11 THEOREM.** *If $\alpha > -1$ and $h > 0$ then $(C,\alpha) \subset (C,\alpha+h)$, i.e. if the sequence $\{z_k\}_0^{\infty}$ is $(C,\alpha)$-summable to $y$ then $\{z_k\}_0^{\infty}$ is $(C,\alpha+h)$-summable to $y$ also.*

*Proof.* Assume that the sequence $\{z_k\}^{\infty}$ is $(C,\alpha)$-summable to $y$. i.e. $\lim_{k\to\infty} \sum_{j=0}^{k} c_{k,\,j}^{\,\alpha} z_j = y$.  Let

$$t_k = \sum_{j=0}^{k} c_{k,\,j}^{\,\alpha} z_j \quad \text{and} \quad \sigma_k = \sum_{j=0}^{k} c_{k,\,j}^{\,\alpha+h} z_j.$$

Thus,

$$A_k^{\alpha} t_k = A_k^{\alpha} \sum_{j=0}^{k} c_{k,\,j}^{\,\alpha} z_j \equiv \sum_{j=0}^{k} A_k^{\alpha} c_{k,\,j}^{\,\alpha} z_j = \sum_{j=0}^{k} d_{k,\,j}^{\,\alpha} z_j$$

and hence,

$$\sum_{k=0}^{t} d_{t,\,k}^{-\alpha} A_k^{\alpha} t_k = \sum_{k=0}^{t} d_{t,\,k}^{-\alpha} (\sum_{j=0}^{k} d_{k,\,j}^{\,\alpha} z_j)$$

$$= \sum_{j=0}^{t} (\sum_{k=j}^{t} d_{t,\,k}^{-\alpha} d_{k,\,j}^{\,\alpha}) z_j = z_t.$$

Therefore,

$$\sigma_k = \sum_{j=0}^{k} c_{k,\,j}^{\,\alpha+h} z_j = \sum_{j=0}^{k} c_{k,\,j}^{\,\alpha+h} (\sum_{m=0}^{j} d_{j,\,m}^{-\alpha} A_m^{\alpha} t_m)$$

$$= \sum_{m=0}^{k} (\sum_{j=m}^{k} c_{k,\,j}^{\,\alpha+h} d_{j,\,m}^{-\alpha} A_m^{\alpha}) t_m.$$

Now, if the transform $(p_{k,m})$, where $p_{k,m} = 0$ if $k < m$ and

$$p_{k,m} = \sum_{j=m}^{k} c_{k,j}^{\alpha+h} \, d_{j,m}^{-\alpha} \, A_m^{\alpha} \quad \text{if } k \geqslant m,$$ is regular then the sequence $\{\sigma_k\}_0^{\infty}$

converges to $y$ (since the sequence $\{t_k\}_0^{\infty}$ converges to $y$) and the theorem will be proved.

First, $p_{m,m} = A_m^{\alpha}/A_m^{\alpha+h}$ and, if $k > m$ then

$$p_{k,m} = \sum_{j=m}^{k} c_{k,j}^{\alpha+h} \, d_{j,m}^{-\alpha} \, A_m^{\alpha} = \frac{A_m^{\alpha}}{A_k^{\alpha+h}} \sum_{j=m}^{k} d_{k,j}^{\alpha+h} \, d_{j,m}^{-\alpha} \quad (3.2)$$

$$= \frac{A_m^{\alpha}}{A_k^{\alpha+h}} \frac{(\alpha+h-\alpha) \ldots (\alpha+h-\alpha+k-m-1)}{(k-m)!}$$

(using the definition of $d_{k,j}^{\beta}$ and Lemma 3.8). Therefore, if $k > m$ then

$$p_{k,m} = \frac{(\alpha+1) \ldots (\alpha+m)}{m!} \left[ \frac{k-m+1}{\alpha+h+k-m+1} \ldots \frac{k}{\alpha+h+k} \right] \quad (3.3)$$

$$\left[ \frac{h}{\alpha+h+1} \ldots \frac{h+k-m-1}{\alpha+h+k-m} \right].$$

Let $m$ be fixed and take $k > m+1$. We have that

$$\frac{k-m+1}{\alpha+h+k-m+1} \ldots \frac{k}{\alpha+h+k} \sim \frac{k^m}{k^m}, \quad (k \to \infty)$$

so $\lim_{k \to \infty} \left[ \frac{k-m+1}{\alpha+h+k-m+1} \ldots \frac{k}{\alpha+h+k} \right] = 1.$

Also,

$$\frac{h}{\alpha+h+1} \ldots \frac{h+k-m-1}{\alpha+h+k-m} = \left( 1 - \frac{\alpha+1}{\alpha+h+1} \right) \ldots \left( 1 - \frac{\alpha+1}{\alpha+h+k-m} \right)$$

and $0 \leqslant b_j < 1$ for all $j = 1, 2, \ldots$ where $b_j = (\alpha+1)/(\alpha+h+j-m)$. So, by the exercise preceding the statement of this theorem and the fact that $\sum_{j=1}^{\infty} b_j$ diverges, we have that

$$\lim_{k \to \infty} \left[ \frac{h}{\alpha+h+1} \ldots \frac{h+k-m-1}{\alpha+h+k-m} \right] = 0.$$

So, by Eq. (3.3), $\lim_{k \to \infty} p_{k,m} = 0$ (thus, condition (i) of Theorem 2.3 holds).

From Eq. (3.3) we have that $p_{k,m} \geqslant 0$.     Therefore, if $\lim_{k \to \infty} \sum_{m=0}^{\infty} p_{k,m} = 1$ then conditions (ii) and (iii) of Theorem 2.3 will hold and, hence $(p_{k,m})$ will be regular.   Examine

$$\sum_{m=0}^{\infty} p_{k,m} = \sum_{m=0}^{k} p_{k,m} = \sum_{m=0}^{k} \frac{A_m^{\alpha}}{A_k^{\alpha+h}} \sum_{j=m}^{k} d_{k,j}^{\alpha+h} d_{j,m}^{-\alpha}$$

$$= \frac{1}{A_k^{\alpha+h}} \sum_{m=0}^{k} A_m^{\alpha} \frac{h \ldots (h+k-m-1)}{(k-m)!}$$

(by Eq. (1)).   From the proof of Lemma 3.8 we have

$$\sum_{m=0}^{k} A_m^{\alpha} \frac{h \ldots (h+k-m-1)}{(k-m)!} = \sum_{m=0}^{k} \frac{(\alpha+1) \ldots (\alpha+m)}{m!} \frac{h \ldots (h+k-m-1)}{(k-m)!} = c_k$$

where

$$1 + \sum_{k=1}^{\infty} c_k x^k = [1 + \sum_{j=1}^{\infty} \frac{(\alpha+1) \ldots (\alpha+j)}{j!} x^j][1 + \sum_{t=1}^{\infty} \frac{h \ldots (h+t-1)}{t!} x^t]$$

$$= (1-x)^{-(\alpha+1)} (1-x)^{-h} = (1-x)^{-(\alpha+h+1)}$$

$$= 1 + \sum_{k=1}^{\infty} \frac{(\alpha+h+1) \ldots (\alpha+h+k)}{k!} x^k,$$

i.e.

$$\sum_{m=0}^{k} A_m^{\alpha} \frac{h \ldots (h+k-m-1)}{(k-m)!} = \frac{(\alpha+h+1) \ldots (\alpha+h+k)}{k!} = A_k^{\alpha+h}$$

Therefore,

$$\sum_{m=0}^{k} p_{k,m} = \frac{1}{A_k^{\alpha+h}} \sum_{m=0}^{k} A_m^{\alpha} \frac{h \ldots (h+k-m-1)}{(k-m)!} = 1. \quad \triangle$$

### Euler, Taylor and Borel exponential transformations

Next, we examine three methods of summability (Euler, Taylor and Borel exponential), two of which are functions of a parameter and the third (the Borel exponential method) is a so-called 'continuous', method of summability.   We begin with the Borel exponential method.

3.12  DEFINITION.  *The sequence* $\{z_k\}_0^\infty$ *is Borel summable to y (written (B)-summable to y) if*

$$\sum_{k=0}^\infty z_k \frac{x^k}{k!}$$

*converges for all real x and*

$$\lim_{x \to \infty} e^{-x} \sum_{k=0}^\infty z_k \frac{x^k}{k!} = y \quad exists.$$

E. Borel first defined the Borel exponential method in the article 'Mémoire sur les séries divergentes', *Annls. scient. Éc. norm. sup.* Paris (3), volume 16, pages 9-136 in 1899. G.H. Hardy (in 1903) improved and corrected some of Borel's results in an article in *Trans. Camb. Phil. Soc.*, volume 19, pages 297-321 and in 'Researches in the theory of divergent series and divergent integrals', *Q. J. Math.*, volume 35, pages 22–66.

3.13  THEOREM.  *The Borel exponential method is regular.*

*Proof.*  Let $\{z_k\}_0^\infty$  converge  (say  to  $z$).      Therefore, $\lim_{n \to \infty} \sup |z_n/n!|^{1/n} = 0$ and, hence, the series $\sum_{k=0}^\infty z_k (x^k/k!)$ converges for all real $x$.

Now, let $\varepsilon > 0$ be given and consider

$$\left| e^{-x} \sum_{k=0}^\infty z_k \frac{x^k}{k!} - z \right| - \left| e^{-x} \sum_{k=0}^\infty z_k \frac{x^k}{k!} - e^{-x} \sum_{k=0}^\infty \frac{x^k}{k!} z \right|$$

$$= \left| e^{-x} \sum_{k=0}^\infty (z_k - z) \frac{x^k}{k!} \right| .$$

There exists a positive integer $K = K(\varepsilon)$ such that if $k \geqslant K$ then $|z_k - z| < \frac{1}{2}\varepsilon$. Also, since $\{z_k\}_0^\infty$ converges there exists $M > 0$ such that $|z_k - z| \leqslant M$ for all $k = 0, 1, \ldots$. Therefore,

$$\left| e^{-x} \sum_{k=0}^\infty z_k \frac{x^k}{k!} - z \right| \leqslant \left| e^{-x} \sum_{k=0}^{K-1} (z_k - z) \frac{x^k}{k!} \right|$$

$$+ \left| e^{-x} \sum_{k=K}^\infty (z_k - z) \frac{x^k}{k!} \right|$$

$$< e^{-x} M K x^{K-1} + \left( e^{-x} \sum_{k=0}^\infty \frac{x^k}{k!} \right) \frac{1}{2}\varepsilon$$

$$= e^{-x} M K x^{K-1} + \frac{1}{2}\varepsilon \quad \text{for } x \geqslant 1.$$

Choose $\delta = \delta(\varepsilon) > 1$ such that if $x \geqslant \delta$ then $\dfrac{x^{K-1}}{e^x} < \dfrac{\varepsilon}{2MK}$.   So, for $x > \delta$ we have

$$\left| e^{-x} \sum_{k=0}^{\infty} z_k \frac{x^k}{k!} - z \right| < MK \frac{\varepsilon}{2MK} + \frac{1}{2}\varepsilon = \varepsilon,$$

i.e.

$$\lim_{x \to \infty} e^{-x} \sum_{k=0}^{\infty} z_k \frac{x^k}{k!} = z$$

and, hence, the Borel exponential method is regular.   $\triangle$

Frequently the Borel exponential method is replaced by a matrix method (called the Borel matrix method of summability).  To do this let

$$b_{n,k} = e^{-n} \frac{n^k}{k!} \quad \text{for all } n = 0, 1, \ldots \text{ and } k = 0, 1, \ldots .$$

Define the Borel transform of the sequence $\{z_k\}_0^{\infty}$ to be the sequence $\{\sigma_n\}_0^{\infty}$ where

$$\sigma_n = \sum_{k=0}^{\infty} b_{n,k} z_k.$$

*Exercise.* (i) Prove that the Borel matrix method of summability is regular.

(ii) Prove that the Borel exponential method is included in the Borel matrix method.

To see that the Borel exponential method and the Borel matrix method are not mutually consistent over the entire space of sequences let $f(x) = e^x \sin \pi x$ and construct the Maclaurin series $\sum_{k=0}^{\infty} a_k x^k$ which converges to $f(x)$ for all real $x$ (such a series exists).  Define the sequence $\{z_k\}_0^{\infty}$ by $z_k = k! a_k$.   Clearly the sequence $\{z_k\}_0^{\infty}$ is summable to 0 by the Borel matrix method of summability, however,

$$\lim_{x \to \infty} e^{-x} \sum_{k=0}^{\infty} z_k \frac{x^k}{k!} = \lim_{x \to \infty} \sin \pi x$$

does not exist.

Much of the work on the Euler transform of order $r$ (see Definition 3.14) was done by K. Knopp in articles 'Über das Eulersche Summierungsverfahren', I and II, *Math. Z.*, volume 15, pages 226–53 in 1922 and volume 16, pages 125–56 in 1923.  The original Euler transformation (when $r = \frac{1}{2}$) was given by L. Euler in *Institutiones calculi differentialis*, page 281 in 1755.

3.14 DEFINITION. *Let* $r \in \mathbb{C} \setminus \{1, 0\}$. *The Euler transform of order* $r$ *(($E, r$)-transform) is the matrix* $(e_{n,k}^r)$ *where*

$$e_{n,k}^{r} = \begin{cases} \binom{n}{k} r^k (1-r)^{n-k} & \text{if } k \leqslant n \\ \\ 0 & \text{if } k > n. \end{cases} \qquad \text{For } r = 1 \text{ let } e_{n,k}^{1} = \begin{cases} 0 & \text{if } k \neq n \\ \\ 1 & \text{if } k = n \end{cases}$$

and $e_{n,k}^{0} = 0$ for $n = 0, 1, \ldots$ ; $k \geqslant 1$ and $e_{n,0}^{0} = 1$ for $n = 0, 1, \ldots$ .

Euler's original transformation, $E_1$, was given by the matrix $(a_{n,k})$ where

$$a_{n,k} = \begin{cases} \binom{n}{k} \frac{1}{2^n} & \text{if } k \leqslant n \\ \\ 0 & \text{if } k > n. \end{cases}$$

The original generalization of the $E_1$ transform was the transform $(b_{n,k}^{q})$ where

$$b_{n,k}^{q} = \begin{cases} \binom{n}{k} (q+1)^{-n} q^{n-k} & \text{if } k \leqslant n \\ \\ 0 & \text{if } k > n. \end{cases}$$

Thus, $(b_{n,k}^{1})$ gave the original $E_1$ transform.

In Definition 3.14 if we let $r = (q+1)^{-1}$ we obtain the $(b_{n,k}^{q})$-transform. Therefore, the original $E_1$ transform is the $(E, \frac{1}{2})$-transform, i.e. when $r = \frac{1}{2}$.

3.15 THEOREM. *The $(E,r)$-transform is regular if and only if $0 < r \leqslant 1$, i.e. $r$ is real and $0 < r \leqslant 1$.*

*Proof.* We have that the series $\sum\limits_{n=k}^{\infty} \binom{n}{k} w^{n-k}$ converges for $|w| < 1$ (since the function $(1-w)^{k+1}$ has, as its Taylor series expansion about $w = 0$, the series $\sum\limits_{n=k}^{\infty} \binom{n}{k} w^{n-k}$, and this series converges for $|w| < 1$). Let $k$ be fixed. Thus, $\lim_{n\to\infty} e_{n,k}^{r} = \lim_{n\to\infty} \binom{n}{k} r^k (1-r)^{n-k} = r^k \lim_{n\to\infty} \binom{n}{k}(1-r)^{n-k}$. If $|1-r| \geqslant 1$ then $\lim_{n\to\infty} \binom{n}{k}(1-r)^{n-k} \neq 0$. If $|1-r| < 1$ then we have $\lim_{n\to\infty} \binom{n}{k}(1-r)^{n-k} = 0$ since the series $\sum\limits_{n=k}^{\infty} \binom{n}{k}(1-r)^{n-k}$ converges. Therefore, condition (i) of Theorem 2.3 holds, i.e. $\lim_{n\to\infty} e_{n,k}^{r} = 0$, if and only if $|1-r| < 1$.

Next, since $(\alpha+\beta)^n = \sum\limits_{k=0}^{n} \binom{n}{k} \alpha^k \beta^{n-k}$, we have that

$$\sum_{k=0}^{\infty} e_{n,k}^{r} = \sum_{k=0}^{n} e_{n,k}^{r} = \sum_{k=0}^{n} \binom{n}{k} r^k (1-r)^{n-k} = (1-r+r)^n = 1$$

and, hence, $\lim_{n\to\infty} \sum\limits_{k=0}^{\infty} e_{n,k}^{r} = 1$. Therefore, condition (ii) of Theorem

2.3 holds (without restriction on the parameter $r$).

Finally,

$$\sum_{k=0}^{\infty} |e_{n,k}^{r}| = \sum_{k=0}^{n} \binom{n}{k} |r|^{k} |1-r|^{n-k} = (|r| + |1-r|)^{n}.$$

So, $\sup_{n} \left\{ \sum_{k=0}^{\infty} |e_{n,k}^{r}| \right\} \leqslant M < \infty$ for some $M > 0$ if and only if $\sup_{n} (|r| + |1-r|)^{n} \leqslant M < \infty$, i.e. if and only if $|r| + |1-r| \leqslant 1$. Now, by the triangle inequality, we know that $|r| + |1-r| \geqslant 1$ for all $r \in \mathbb{C}$. So $|r| + |1-r| \leqslant 1$ if and only if $|r| + |1-r| = 1$, i.e. $r$ is a real number and $0 \leqslant r \leqslant 1$. So, condition (iii) of Theorem 2.3 holds if and only if $0 \leqslant r \leqslant 1$. Combining this with the restriction that $|1-r| < 1$ we have that $(e_{n,k}^{r})$ is regular if and only if $r$ is real and $0 < r \leqslant 1$. $\triangle$

We are now going to examine two inclusion theorems involving the Euler transformation and the Borel exponential method of summability $(B)$. Before we do this, however, we need the following lemma, definition, and corollary.

**3.16 LEMMA.** *If $rs \neq 0$ then the product $(e_{n,k}^{r}) (e_{n,k}^{s})$ is the matrix for the $(E,rs)$ transform.*

*Proof.* Let $(a_{n,j}) = (e_{n,k}^{r}) (e_{k,j}^{s})$. Clearly, $a_{n,j} = 0$ if $j > n$. Also, if $r = 1$ or $s = 1$ the result follows trivially (since $(E,1)$ is ordinary convergence). Assume that $r \neq 1$, $s \neq 1$, and that $j \leqslant n$. So $a_{n,j} = \sum_{k=j}^{n} e_{n,k}^{r} e_{k,j}^{s}$. If $j = n$ we have

$$a_{n,n} = e_{n,n}^{r} e_{n,n}^{s} = \binom{n}{n} r^{n} \binom{n}{n} s^{n} = (rs)^{n}.$$

If $j < n$ then

$$\sum_{k=j}^{n} e_{n,k}^{r} e_{j,k}^{s} = \sum_{k=j}^{n} \binom{n}{k} r^{k} (1-r)^{n-k} \binom{k}{j} s^{j} (1-s)^{k-j}$$

$$= s^{j} \frac{(1-r)^{n}}{(1-s)^{j}} \sum_{k=j}^{n} \binom{n}{k} \binom{k}{j} r^{k} \left[\frac{1-s}{1-r}\right]^{k}$$

$$= \binom{n}{j} (rs)^{j} (1-r)^{n-j} \sum_{k=j}^{n} \binom{n-j}{k-j} \left[\frac{r(1-s)}{1-r}\right]^{k-j}$$

$$= \binom{n}{j} (rs)^{j} (1-r)^{n-j} \left[1 + \frac{r(1-s)}{1-r}\right]^{n-j}$$

$$= \binom{n}{j} (rs)^{j} (1-rs)^{n-j}. \qquad \triangle$$

**3.17 COROLLARY.** *If $r \neq 0$ then $(E,r)$ is invertible and*

$(E, r)^{-1} = (E, 1/r)$.

*Proof.* The proof follows directly from Lemma 3.16 by letting $s = 1/r$. $\triangle$

3.18 THEOREM. *If* $0 < |s| \leqslant |r|$ *and* $|s| + |r-s| = |r|$ *then* $(E, r) \subset (E, s)$.

*Proof.* Let $0 < |s| \leqslant |r|$ and suppose that the sequence $\{z_k\}_0^\infty$ is $(E, r)$-summable to $y$, i.e. the sequence $\{t_n\}_0^\infty$ converges to $y$ where

$$t_n = \sum_{k=0}^\infty e_{n,k}^r \, z_k = \sum_{k=0}^n e_{n,k}^r \, z_k. \quad \text{So,}$$

$$\sum_{n=0}^j e_{j,n}^{1/r} \, t_n = \sum_{n=0}^j e_{j,n}^{1/r} \sum_{k=0}^n e_{n,k}^r \, z_k$$

$$= \sum_{k=0}^j \Big( \sum_{n=k}^j e_{j,n}^{1/r} \, e_{n,k}^r \Big) z_k = z_j$$

by Corollary 3.17. Now, let

$$\sigma_n = \sum_{j=0}^\infty e_{k,j}^s \, z_j = \sum_{j=0}^k e_{k,j}^s \, z_j$$

$$= \sum_{j=0}^k e_{k,j}^s \Big( \sum_{n=0}^j e_{j,n}^{1/r} \, t_n \Big)$$

$$= \sum_{n=0}^k \Big( \sum_{j=n}^k e_{k,j}^s \, e_{j,n}^{1/r} \Big) t_n = \sum_{n=0}^k e_{k,n}^{s/r} \, t_n$$

by Lemma 3.16. Since $0 < |s| \leqslant |r|$ and $|s| + |r-s| = |r|$ we have that $0 < s/r \leqslant 1$ and, by Theorem 3.15, the $(E, s/r)$-transform is regular. Therefore, since the sequence $\{t_n\}_0^\infty$ converges to $y$ we have that the sequence $\{\sigma_n\}_0^\infty$ converges to $y$. $\triangle$

3.19 THEOREM. *If* $r > 0$ *then* $(E, r) \subset (B)$.

*Proof.* Assume that $r > 0$ and suppose that the sequence $\{z_k\}_0^\infty$ is $(E, r)$-summable to $y$, i.e. the sequence $\{t_n\}_0^\infty$ converges to $y$ where

$$t_n = \sum_{k=0}^\infty e_{n,k}^r \, z_k = \sum_{k=0}^n e_{n,k}^r \, z_k. \quad \text{By Corollary 3.17 we have that}$$

$$\sum_{n=0}^j e_{j,n}^{1/r} \, t_n = \sum_{n=0}^j e_{j,n}^{1/r} \sum_{k=0}^n e_{n,k}^r \, z_k = \sum_{k=0}^j \Big( \sum_{n=k}^j e_{j,n}^{1/r} \, e_{n,k}^r \Big) z_k = z_j.$$

Now, consider the series

$$\sum_{j=0}^{\infty} z_j \frac{x^j}{j!} = \sum_{j=0}^{\infty} \left( \sum_{n=0}^{j} e_{j,n}^{1/r} t_n \right) \frac{x^j}{j!} = \sum_{n=0}^{\infty} \left( \sum_{j=n}^{\infty} e_{j,n}^{1/r} \frac{x^j}{j!} \right) t_n$$

(provided this sum converges absolutely). Before we examine the condition of absolute convergence let us continue to examine

$$\sum_{n=0}^{\infty} \left( \sum_{j=n}^{\infty} e_{j,n}^{1/r} \frac{x^j}{j!} \right) t_n = \sum_{n=0}^{\infty} \left( \sum_{j=n}^{\infty} \binom{j}{n} (\tfrac{1}{r})^n (1-\tfrac{1}{r})^{j-n} \frac{x^j}{j!} \right) t_n$$

$$= \sum_{n=0}^{\infty} \frac{1}{n!} (\tfrac{x}{r})^n \sum_{j=n}^{\infty} \frac{(x[1-\tfrac{1}{r}])^{j-n}}{(j-n)!} t_n$$

$$= e^{x[1-\frac{1}{r}]} \sum_{n=0}^{\infty} \frac{1}{n!} (\tfrac{x}{r})^n t_n .$$

The double series converges absolutely (and, in particular, converges) for all $x$ since $\{t_n\}_0^{\infty}$ converges. So, the interchange of summation is allowable and, moreover, the original series $\sum_{j=0}^{\infty} z_j \frac{x^j}{j!}$ converges for all $x$. Now,

$$\lim_{x \to \infty} e^{-x} \sum_{j=0}^{\infty} z_j \frac{x^j}{j!} = \lim_{x \to \infty} e^{-x/r} \sum_{n=0}^{\infty} \frac{t_n}{n!} (\tfrac{x}{r})^n = y$$

since the Borel method, $(B)$, is regular and $\{t_n\}_0^{\infty}$ converges to $y$. Therefore, $(E,r) \subset (B)$. $\triangle$

Now we turn our attention to a third transform, namely, the Taylor transform $T(r)$.

3.20 DEFINITION. *Let* $r \in \mathbb{C} \setminus \{0\}$. *The Taylor transform* $T(r) = (c_{n,k}^{r})$ *is defined by*

$$c_{n,k} = \begin{cases} 0 & \text{if } k < n \\ \binom{k}{n} r^{k-n} (1-r)^{n+1} & \text{if } k \geqslant n. \end{cases}$$

*For* $r = 0$ *let*
$$c_{n,k} = \begin{cases} 0 & \text{if } k \neq n \\ 1 & \text{if } k = n. \end{cases}$$

In 1916 G.H. Hardy and J.E. Littlewood introduced the Taylor method (known then as the circle method of order $r$) for $0 < r < 1$ in the article 'Theorems Concerning the Summability of Series by Borel's Exponential Method', *Rc. Circ. Mat. Palermo*, volume 41, pages 36–53.

The name of Taylor became associated with the transform because the original work depended largely upon the use of Taylor series expansions.

In working with the Taylor transform we will use, frequently, the fact that $(1-w)^{-(n+1)} = \sum\limits_{k=n}^{\infty} \binom{k}{n} w^{k-n}$ if $|w| < 1$ and, moreover, $\sum\limits_{k=n}^{\infty} \binom{k}{n} w^{k-n}$ converges only if $|w| < 1$. First, we find restrictions on $r$ such that $T(r)$ is regular.

3.21  THEOREM. *The Taylor transform,* $T(r)$, *is regular if and only if* $0 \leqslant r < 1$, *i.e.* $r$ *is real and* $0 \leqslant r < 1$.

*Proof.* If $r = 1$ then the Taylor transform becomes the matrix with all 0 entries and, hence, transforms all sequences into the sequence consisting of all zeros. So, assume that $r \neq 1$. Let $k$ be fixed and consider $\lim_{n\to\infty} c_{n,k}^{r}$. We have that this limit is 0 since $c_{n,k}^{r} = 0$ for $n > k$. So, condition (i) of Theorem 2.3 is satisfied (without restriction on $r$). Next, consider

$$\sum_{k=0}^{\infty} c_{n,k}^{r} = \sum_{k=n}^{\infty} c_{n,k}^{r} = \sum_{k=n}^{\infty} \binom{k}{n} r^{k-n} (1-r)^{n+1}$$

$$= (1-r)^{n+1} \sum_{k=n}^{\infty} \binom{k}{n} r^{k-n}$$

which converges only if $|r| < 1$ and, in this case,

$$(1-r)^{n+1} \sum_{k=n}^{\infty} \binom{k}{n} r^{k-n} = (1-r)^{n+1} (1-r)^{-(n+1)} = 1.$$

Hence, $\lim_{n\to\infty} \sum\limits_{k=0}^{\infty} c_{n,k}^{r} = 1$ if and only if $|r| < 1$ (condition (ii) of Theorem 2.3). Finally, we have

$$\sum_{k=0}^{\infty} |c_{n,k}^{r}| = \sum_{k=n}^{\infty} |c_{n,k}^{r}| = \sum_{k=n}^{\infty} |1-r|^{n+1} \binom{k}{n} |r|^{k-n} = \frac{|1-r|^{n+1}}{(1-|r|)^{n+1}}$$

since $|r| < 1$. Therefore, $\sup_n \{ \sum\limits_{k=0}^{\infty} |c_{n,k}^{r}| \}$ is finite if and only if $|1-r|/(1-|r|) \leqslant 1$, i.e. $|1-r| \leqslant 1-|r|$. By the triangle inequality we have that $|1-r| \geqslant 1 - |r|$ so $\sup_n \{ \sum\limits_{k=0}^{\infty} |c_{n,k}^{r}| \}$ is finite if and only if $|1-r| = 1 - |r|$, i.e. $r$ is real and $0 \leqslant r \leqslant 1$ (which takes care of condition (iii) of Theorem 2.3). Therefore, $T(r)$ is regular if and only if $|r| < 1$ and $0 \leqslant r \leqslant 1$, i.e. $0 \leqslant r < 1$.  $\triangle$

3.22  DEFINITION. *Let* $(a_{n,k})$ *be an infinite matrix.*

(i)  If $K \in \mathbb{C}$ then $K(a_{n,k})$ is the matrix $(b_{n,k})$ where $b_{n,k} = Ka_{n,k}$.

(ii)  The transpose of the matrix $(a_{n,k})$ (written $(a_{n,k})'$) is the matrix $(b_{n,k})$ where $b_{n,k} = a_{k,n}$.

**3.23  THEOREM.**  *The product of the $T(r)$ matrix with the $T(s)$ matrix is the matrix $(1-r)(1-s)(E,(1-r)(1-s))'$.*

*Proof.*  Let $(d_{n,k}) = (c_{n,j})(c_{j,k})$, i.e. $d_{n,k} = 0$ if $k < n$ and, if $k \geqslant n$ then

$$d_{n,k} = \sum_{j=n}^{k} c^r_{n,j}\, c^s_{j,k} = \sum_{j=n}^{k} \binom{j}{n} r^{j-n} (1-r)^{n+1} \binom{k}{j} s^{k-j} (1-s)^{j+1}$$

$$= \binom{k}{n} [(1-r)(1-s)]^{n+1} s^{k-n} \sum_{j=n}^{k} \binom{k-n}{j-n} \left(\frac{r(1-s)}{s}\right)^{j-n}$$

$$= \binom{k}{n} [(1-r)(1-s)]^{n+1} s^{k-n} \left(1 + \frac{r(1-s)}{s}\right)^{k-n}$$

$$= \binom{k}{n} [(1-r)(1-s)]^{n+1} (s + r - rs)^{k-n}.$$

Now, by Definition 3.14, we have (for $t = (1-r)(1-s)$ and $k \geqslant n$) that

$$e^t_{k,n} = \binom{k}{n} [(1-r)(1-s)]^n (1 - (1-r)(1-s))^{k-n}$$

$$= \binom{k}{n} [(1-r)(1-s)]^n (r + s - rs)^{k-n}. \quad \triangle$$

**3.34  THEOREM.**  *If $r \neq 1$ then $T(r)$ is invertible and $T(r)^{-1} = T[-r/(1-r)]$.*

*Proof.*  Let $s = -r/(1-r)$ in Theorem 3.23.  We have that

$$(1-r)(1-s)(E,(1-r)(1-s))' = (1-r)(1 + \frac{r}{1-r})(E,(1-r)(1 + \frac{r}{1-r}))'$$

$$= (E,1)' = (E,1).$$

The result follows since $(E,1)$ is ordinary convergence.  $\triangle$

**3.25  THEOREM.**  *If  (i) $|r| < \frac{1}{2}$,*
*(ii) $|s-r| = |1-r| - |1-s|$,  and*
*(iii) $|r| < |s|$*
*then any bounded sequence that is $T(r)$ summable is $T(s)$ summable*

*to the same value.*

*Proof.* Let $T(r) = (c_{n,k}^r)$, $T(s) = (c_{n,k}^s)$, and $T[-r/(1-r)] = (q_{n,k})$.

Assume that the bounded sequence $\{z_n\}_0^\infty$ is $T(r)$-summable to $y$, i.e. the sequence $\{\sigma_n\}_0^\infty$ converges to $y$ where

$$\sigma_n = \sum_{k=n}^\infty c_{n,k}^r z_k.$$

Let $\tau_j = \sum_{k=j}^\infty q_{j,k} \sigma_k$.  Thus

$$\tau_j = \sum_{k=j}^\infty q_{j,k} \left( \sum_{n=k}^\infty c_{k,n}^r z_n \right)$$

$$= \sum_{n=j}^\infty \left( \sum_{k=j}^n q_{j,k} c_{k,n}^r \right) z_n$$

$$= z_j \tag{3.4}$$

by Theorem 3.24 provided the interchange of the order of summation is valid.  To show this, examine the absolute convergence of the series in (3.4).  We have

$$S = \sum_{k=j}^\infty |q_{j,k}| \left( \sum_{n=k}^\infty |c_{k,n}| \, |z_n| \right)$$

$$\leqslant M \sum_{n=j}^\infty \sum_{k=j}^n |q_{j,k}| \, |c_{k,n}^r|$$

where $|z_n| \leqslant M$ for all $n = 0, 1, \ldots$ .    So

$$S \leqslant M \sum_{n=j}^\infty \sum_{k=j}^\infty \left| (1-r)^{-(j+1)} \binom{k}{j} (-1)^{k-j} \left(\frac{r}{1-r}\right)^{k-j} \right| \, \left| (1-r)^{k+1} \binom{n}{r} r^{n-k} \right|$$

$$= M \sum_{n=j}^\infty \binom{n}{j} |r|^{n-j} \sum_{k=j}^n \binom{n-j}{n-k}$$

$$= M \sum_{n=j}^\infty \binom{n}{j} (2|r|)^{n-j}$$

$$= \frac{M}{1-2|r|}$$

since $|r| < \frac{1}{2}$ (condition (i)).


Now, let $\{t_n\}_0^\infty$ be the $T(s)$-transform of the sequence $\{z_k\}_0^\infty$, i.e.

$$t_n = \sum_{k=n}^{\infty} c_{n,k}^{\ s} \, z_k$$

$$= \sum_{k=n}^{\infty} c_{n,k}^{\ s} \, \tau_k$$

$$= \sum_{k=n}^{\infty} c_{n,k}^{\ s} \left( \sum_{j=k}^{\infty} q_{k,j} \, \sigma_j \right).$$

So

$$t_n = \sum_{k=n}^{\infty} \sum_{j=k}^{\infty} (1-s)^{n+1} \binom{k}{n} s^{k-n} (1-r)^{-(k+1)} \binom{j}{k} (-1)^{j-k} \left( \frac{r}{1-r} \right)^{j-k} \sigma_j. \quad (3.5)$$

We now interchange the order of summation (after which we will remark on the absolute convergence of the series in (3.5)). We have, by Eq. (3.5),

$$t_n = (1-s)^{n+1} \sum_{j=n}^{\infty} \binom{j}{n} (-r)^{j-n} (1-r)^{-(j+1)} \left[ \sum_{k=n}^{j} \binom{j-n}{k-n} \left( -\frac{s}{r} \right)^{k-n} \right] \sigma_j$$

$$= (1-s)^{n+1} \sum_{j=n}^{\infty} \binom{j}{n} (-r)^{j-n} (1-r)^{-(j+1)} \left( 1 - \frac{s}{r} \right)^{j-n} \sigma_j.$$

So

$$t_n = \left[ \frac{1-s}{1-r} \right]^{n+1} \sum_{j=n}^{\infty} \binom{j}{n} \left( \frac{s-r}{1-r} \right)^{j-n} \sigma_j \quad (3.6)$$

which is the $T[(s-r)/(1-r)]$-transform of the sequence $\{\sigma_j\}_0^{\infty}$. By condition (ii), $\arg [(1-s)/(1-r)] = 0$. So $(1-s)/(1-r)$ is a real positive number. Moreover, by condition (iii), $(1-s)/(1-r) < 1$. Therefore, if $R = (s-r)/(1-r)$ then $R = 1 - (1-s)/(1-r)$. So $0 < R < 1$ and, hence, $T(R)$ is regular (Theorem 3.21). Since the sequence $\{\sigma_j\}_0^{\infty}$ converges to $y$ we have, by Eq. (3.6), that the sequence $\{t_n\}_0^{\infty}$ converges to $y$ also, i.e. $T(s)$-transform of the sequence $\{z_k\}_0^{\infty}$ converges to $y$. This proves the theorem provided the interchange of the order of summation in (3.5) is valid. We have,

$$\left[ \frac{1-s}{1-r} \right]^{n+1} \sum_{j=n}^{\infty} \binom{j}{n} \left( \frac{s-r}{1-r} \right)^{j-n} \sigma_j$$

converges absolutely (since $\{\sigma_j\}_0^{\infty}$ is bounded, $(1-s)/(1-r) > 0$, and $0 < (s-r)/(1-r) < 1$). Therefore, since (3.6) is equivalent to (3.5), we have that the series in (3.5) converges absolutely. Hence the rearrangement is valid. $\triangle$

## Hausdorff means

We conclude this chapter by examining Hausdorff means.  To do this we first introduce totally monotone sequences.

3.26   DEFINITION.   *Let* $x = \{x_k\}_0^\infty$ *be a real sequence.   Define*

$$(\Delta^0 x)_n = x_n ,$$

$$(\Delta^1 x)_n = x_n - x_{n+1}$$

*and* $\qquad (\Delta^j x)_n = (\Delta^{j-1} x)_n - (\Delta^{j-1} x)_{n+1} \ (j \geqslant 2).$

*The sequence* $x$ *is totally monotone provided* $(\Delta^j x)_n \geqslant 0$ *for all* $j$, $n = 0, 1, \ldots$ .

We see, for example, that the sequence $x = \{x_k\}^\infty$ where $x_k = 1/(k+1)$ is totally monotone since

$$(\Delta^j x)_n = \int_0^1 x^n (1-x)^{j-1} \, dx - \int_0^1 x^{n+1} (1-x)^{j-1} \, dx$$

$$= \int_0^1 x^n (1-x)^j \, dx \geqslant 0.$$

Also, the sequence $x = \{x_k\}_0^\infty$ where $x_k = r^k$ is totally monotone for $0 \leqslant r \leqslant 1$ since

$$(\Delta^j x)_n = r^n (1-r)^j.$$

3.27   LEMMA.   *If* $x = \{x_k\}_0^\infty$ *is a real sequence then*

$$(\Delta^j x)_n = \sum_{k=0}^j \binom{j}{k} (-1)^k x_{n+k}$$

*Proof.*   The case $j = 0$ holds trivially.   For the case $j = 1$ we have that

$$(\Delta^1 x)_n = x_n - x_{n-1}$$

and

$$\sum_{k=0}^1 \binom{1}{k} (-1)^k x_{n+k} = x_n - x_{n-1}.$$

Suppose that $(\Delta^j x)_n = \sum_{k=0}^j \binom{j}{k} (-1)^k x_{n+k}$ for $j = 0, \ldots, m$, $(m \geqslant 1)$.
We have that

$$(\Delta^{m+1}x)_n = (\Delta^m x)_n - (\Delta^m x)_{n+1}$$

$$= \sum_{k=0}^{m} \binom{m}{k} (-1)^k x_{n+k} - \sum_{k=0}^{m} \binom{m}{k} (-1)^k x_{n+1+k}$$

$$= \sum_{k=0}^{m} \binom{m}{k} (-1)^k x_{n+k} - \sum_{k=1}^{m+1} \binom{m}{k-1} (-1)^{k+1} x_{n+k}$$

$$= x_n + \sum_{k=1}^{m} (-1)^k [\binom{m}{k} + \binom{m}{k-1}] x_{n+k} - (-1)^m x_{n+m+1}$$

$$= x_n + \sum_{k=1}^{m} (-1)^k \binom{m+1}{k} x_{n+k} - (-1)^m x_{n+m+1}$$

$$= \sum_{k=0}^{m+1} \binom{m+1}{k} (-1)^k x_{n+k}. \qquad \triangle$$

*Exercise.* Let $x = \{x_k\}_0^\infty$ be a real sequence. Prove that

$$\sum_{k=0}^{n} \binom{n}{k} (\Delta^{n-k}x)_k = x_0.$$

Now, to lead into the definition of a Hausdorff transformation, we need

**3.28 DEFINITION.** *The matrix $\delta = (\delta_{n,k})$ is defined by*

$$\delta_{n,k} = \begin{cases} (-1)^k \binom{n}{k} & \text{if } k \leqslant n \\ \\ 0 & \text{if } k > n. \end{cases}$$

**3.29 LEMMA.** *The product $\delta\delta$ is the matrix $(e_{n,k})$ where*

$$e_{n,k} = \begin{cases} 1 & \text{if } k = n \\ \\ 0 & \text{otherwise.} \end{cases}$$

The proof follows immediately from Definition 3.28.

**3.30 DEFINITION.** (i) *The matrix $\mu = (\mu_{n,k})$ is a diagonal matrix provided $\mu_{n,k} = 0$ for all $n \neq k$. Let $\mu_{n,n} = \mu_n$ in this case.*

(ii) *If $\mu = (\mu_{n,k})$ is a diagonal matrix then the matrix $u = (u_{n,k})$ defined by $u = \delta\mu\delta = (\delta_{n,m}) (\mu_{m,j}) (\delta_{j,k})$ is a Hausdorff transformation (written $(H, \mu)$).*

Note that the multiplication $\delta\mu\delta$ is well-defined since $\delta$ and $\mu$

are lower triangular matrices.

The class of transformations $\delta\mu\delta$ was originally studied, in 1917, by W. A. Hurwitz and L. L. Silverman in the article on 'The consistency and equivalence of certain definitions of summability', *Trans. Am. math. Soc.*, volume 18, pages 1-20. F. Hausdorff (in 1921) rediscovered this class of transformations and developed the theory by associating the class with the famous 'moment problem for a finite interval' in an article, 'Summationsmethoden und Momentfolgen I', *Math. Z.* volume 9, pages 74-109.

3.31 LEMMA. (i) *If* $(H,\mu) = (u_{n,k})$ *and* $(H,\nu) = (v_{n,k})$ *are any two Hausforff transformation then* $(u_{n,k})(v_{n,k}) = (v_{n,k})(u_{n,k})$, i.e. *any two Hausdorff transformations commute.*

(ii) *The product of two Hausdorff transformations is a Hausdorff transformation.*

*Proof.* Let $\delta\mu\delta = (u_{n,k})$ and $\delta\nu\delta = (v_{n,k})$. Thus

$$(\delta\mu\delta)(\delta\nu\delta) = (\delta\mu)(\delta\delta)(\nu\delta) = (\delta\mu)(\nu\delta) = \delta(\mu\nu)\delta = \delta(\nu\mu)\delta$$

$$= (\delta\nu)(\delta\delta)(\mu\delta) = (\delta\nu\delta)(\delta\mu\delta) \text{ provided } \mu\nu = \nu\mu.$$

Let $\mu = (\mu_{n,k})$ and $\nu = (\nu_{n,k})$ be diagonal matrices. Let $(d_{n,k}) = (\mu_{n,k})(\nu_{n,k})$. Thus, $d_{n,k} = 0$ if $n \neq k$ and, if $n = k$ then $d_{n,k} = \mu_n \nu_n$. Therefore, it follows that $\mu\nu = \nu\mu$. We also have that the product $(H,\mu)(H,\nu)$ is merely the Hausdorff transformation $(H,\mu\nu)$. $\triangle$

Theorem 3.31 yields

3.32 COROLLARY. *The Hausdorff matrix* $(H,\mu)$ *is invertible provided* $\mu_n \neq 0$ $(n = 0, 1, \ldots)$. *In this case* $(H,\mu)^{-1} = (H,\nu)$ *where* $\nu_n = 1/\mu_n$ (*call* $\nu = 1/\mu$).

Examples of Hausdorff transformations are the $(C,1)$ transformation [let $\mu_n = 1/(n+1)$] and the Euler transformation (let $\mu_n = r^n$).

An interesting classification theorem for Hausdorff matrices is that the class of Hausdorff matrices is the collection of all lower triangular matrices which commute with the $(C,1)$ matrix. This result is Corollary 3.34.

3.33 LEMMA. *Let* $(H,\mu)$ *be a Hausdorff matrix where* $\mu_n \neq \mu_m$ *for* $m \neq n$. *If the matrix* A *is a lower triangular matrix that commutes with* $(H,\mu)$ *then* A *is a Hausdorff matrix.*

*Proof.* Let $A = (a_{n,k})$ be a lower triangular matrix that commutes with the matrix $\lambda = \delta\mu\delta$. Define the matrix $\omega = \delta A\delta$ (so $A = \delta\omega\delta$ by Lemma 3.29). If $\omega$ is a diagonal matrix then $A$ is a Hausdorff transformation. Consider

$$\omega\mu = (\delta A\delta)(\delta\lambda\delta) = \delta A\lambda\delta$$

and

$$\mu\omega = (\delta\lambda\delta)(\delta A\delta) = \delta\lambda A\delta.$$

Since $A\lambda = \lambda A$ we have that $\omega\mu = \mu\omega$. Suppose that $\omega = (\omega_{n,k})$. Thus

$$\omega\mu = (\ell_{n,k}) \quad \text{where } \ell_{n,k} = \begin{cases} 0 & \text{if } k > n \\ \\ \omega_{n,k}\mu_k & \text{if } k \leqslant n \end{cases}$$

and

$$\mu\omega = (r_{n,k}) \quad \text{where } r_{n,k} = \begin{cases} 0 & \text{if } k > n \\ \\ \mu_n\omega_{n,k} & \text{if } k \leqslant n. \end{cases}$$

Thus $\ell_{n,k} = r_{n,k}$ if and only if

$$\omega_{n,k}\mu_k = \mu_n\omega_{n,k} \quad \text{for all } k \leqslant n.$$

If $k = n$ this holds. If $k < n$ then we have

$$\omega_{n,k}(\mu_k - \mu_n) = 0$$

hence $\omega_{n,k} = 0$ (since $\mu_k \neq \mu_n$ for $k \neq n$). Therefore, $\omega$ is a diagonal matrix. $\triangle$

**3.34 COROLLARY.** *A matrix $A$ is a Hausdorff transformation if and only if it commutes with the $(C;1)$ transformation.*

    The proof follows directly from Lemma 3.33 and the fact that the $(C,1)$ transformation is a Hausdorff transformation with diagonal elements differing.

**3.35 LEMMA.** *If $(H,\mu) = (h_{n,k})$ is a Hausdorff transformation then*

$$h_{n,k} = \begin{cases} 0 & \text{if } k > n \\ \\ \binom{n}{k}(\Delta^{n-k}\mu)_k & \text{if } k \leqslant n. \end{cases}$$

*Proof.* We have that

$$h_{n,k} = \begin{cases} 0 & \text{if } k > n \\ \\ \sum\limits_{j=k}^{n} \binom{n}{j} \binom{j}{k} (-1)^{j+k} \mu_j & \text{if } k \leqslant n. \end{cases}$$

Now
$$\sum_{j=k}^{n} \binom{n}{k} \binom{j}{k} (-1)^{j+k} \mu_j = \binom{n}{k} \sum_{j=k}^{n} \binom{n-k}{j-k} (-1)^{j+k} \mu_j$$

$$= \binom{n}{k} \sum_{j=0}^{n-k} \binom{n-k}{j} (-1)^{j+2k} \mu_{j+k}$$

$$= \binom{n}{k} \sum_{j=0}^{n-k} \binom{n-k}{j} (-1)^{j} \mu_{j+k}$$

$$= \binom{n}{k} (\Delta^{n-k} \mu)_k \quad \text{by Lemma 3.27.} \quad \triangle$$

This preceeding lemma allows us to pursue conditions on $\mu$ such that $(H,\mu)$ is regular (see Theorem 3.39). The next three lemmas are preparatory to proving Theorem 3.39.

3.36 LEMMA. *If $(H,\mu) = (h_{n,k})$ is a Hausdorff matrix then $\sum\limits_{k=0}^{n} h_{n,k} = \mu_0$ for all $n = 0, 1, \ldots$ .*

*Proof.* We have that

$$\sum_{k=0}^{n} h_{n,k} = \sum_{k=0}^{n} \sum_{j=k}^{n} \binom{n}{j} \binom{j}{k} (-1)^{j+k} \mu_j$$

$$= \sum_{j=0}^{n} \binom{n}{j} (-1)^{j} \left[ \sum_{k=0}^{j} \binom{j}{k} (-1)^{k} \right] \mu_j$$

$$= \mu_0 + \sum_{j=1}^{n} \binom{n}{j} (-1)^{j} \left[ \sum_{k=0}^{j} \binom{j}{k} (-1)^{k} \right] \mu_j$$

$$= \mu_0 . \quad \triangle$$

3.37 LEMMA. *The following statements are equivalent:*
(i) *There exist totally monotone sequences $\alpha = \{\alpha_n\}_0^{\infty}$ and $\beta = \{\beta_n\}^{\infty}$ such that $\mu_n = \alpha_n - \beta_n$.*
(ii) *There exists $M > 0$ such that*
$$\sup_n \left\{ \sum_{k=0}^{n} \binom{n}{k} \mid (\Delta^{n-k}\mu)_k \mid \right\} \leqslant M.$$

*Proof.* First we will show that (i) implies (ii). Assume that $\mu_n = \alpha_n - \beta_n$ where $\{\alpha_n\}_0^\infty$ and $\{\beta_n\}_0^\infty$ are totally monotone sequences. Therefore

$$\sum_{k=0}^n \binom{n}{k} \mid (\Delta^{n-k}\mu)_k \mid = \sum_{k=0}^n \binom{n}{k} \mid \sum_{j=0}^{n-k} \binom{n-k}{j} (-1)^j \mu_{k+j} \mid$$

$$= \sum_{k=0}^n \binom{n}{k} \mid \sum_{j=0}^{n-k} \binom{n-k}{j} (-1)^j (\alpha_{k+j} - \beta_{k+j}) \mid$$

$$= \sum_{k=0}^n \binom{n}{k} \mid \sum_{j=0}^{n-k} \binom{n-k}{j} (-1)^j \alpha_{k+j}$$

$$- \sum_{j=0}^{n-k} \binom{n-k}{j} (-1)^j \beta_{k+j} \mid$$

$$= \sum_{k=0}^n \binom{n}{k} \mid (\Delta^{n-k}\alpha)_k - (\Delta^{n-k}\beta)_k \mid$$

$$\leqslant \sum_{k=0}^n \binom{n}{k} (\Delta^{n-k}\alpha)_k + \sum_{k=0}^n \binom{n}{k} (\Delta^{n-k}\beta)_k$$

since $\alpha$ and $\beta$ are totally monotone. Now

$$\sum_{k=0}^n \binom{n}{k} (\Delta^{n-k}\alpha)_k = \alpha_0 \quad \text{and} \quad \sum_{k=0}^n \binom{n}{k} (\Delta^{n-k}\beta)_k = \beta_0$$

thus (i) implies (ii).

To prove that (ii) implies (i) assume that there exists $M > 0$ such that

$$\sup_n \{ \sum_{k=0}^n \binom{n}{k} \mid (\Delta^{n-k}\mu)_k \mid \} \leqslant M.$$

Now,

$$\mid (\Delta^m \mu)_p \mid = \mid (\Delta^m \mu)_p - (\Delta^m \mu)_{p+1} + (\Delta^m \mu)_{p+1} \mid$$

$$\leqslant \mid (\Delta^{m+1}\mu)_p \mid + \mid (\Delta^m \mu)_{p+1} \mid.$$

Continuing this argument, we have that

$$\mid (\Delta^m \mu)_p \mid \leqslant \sum_{j=0}^k \binom{k}{j} \mid (\Delta^{m+k-j}\mu)_{p+j} \mid.$$

Define

$$D_{m,\,p}^{(k)} = \sum_{j=0}^k \binom{k}{j} \mid (\Delta^{m+k-j}\mu)_{p+j} \mid$$

and define

$$M_n = \sum_{k=0}^n \binom{n}{k} \mid (\Delta^{n-k}\mu)_k \mid.$$

We have that $D_{0,\,0}^{(k)} = M_k$ and $D_{m,\,p}^{(k)} \leqslant D_{m,\,p}^{(k+1)}$ (for all $m$ and $p$) hence $\{M_k\}_0^\infty$ is a non-decreasing sequence bounded above (by $M$) and is, thus, convergent. Furthermore,

$$
\begin{aligned}
D_{m,\,p}^{(k)} &= \sum_{j=0}^{k} \binom{k}{j} \mid (\Delta^{m+k-j}\mu)_{p+j} \mid \\
&\leqslant \sum_{j=0}^{k} \binom{m+p+k}{p+j} \mid (\Delta^{m+k-j}\mu)_{p+j} \mid \\
&= \sum_{j=0}^{p+k} \binom{m+p+k}{j} \mid (\Delta^{m+p+k-j}\mu)_j \mid = M_{m+p+k} \leqslant M
\end{aligned}
$$

thus $\{D_{m,\,p}^{(k)}\}_{k=0}^\infty$ is a non-decreasing sequence bounded above. Define

$$
\sigma_{m,p} = \lim_{k \to \infty} D_{m,\,p}^{(k)} \geqslant 0
$$

and note that

$$
D_{m,\,p}^{(k)} \leqslant D_{m+1,\,p}^{(k)} + D_{m,p+1}^{(k)} \leqslant D_{m,\,p}^{(k+1)}.
$$

Therefore, $\sigma_{m,p} \leqslant \sigma_{m+1,p} + \sigma_{m,p+1} \leqslant \sigma_{m,p}$, i.e.

$$
\sigma_{m+1,p} = \sigma_{m,p} - \sigma_{m,p+1}.
$$

Define the sequence $\{\sigma_p\}_0^\infty$ by $\sigma_p = \sigma_{0,p}$. Since

$$
(\Delta^m \sigma)_p = \sigma_{m,p} \geqslant 0
$$

we have that the sequence $\{\sigma_p\}_0^\infty$ is totally monotone. Define the sequences $\alpha = \{\alpha_p\}_0^\infty$ and $\beta = \{\beta_p\}_0^\infty$ by

$$
\alpha_p = \tfrac{1}{2}(\sigma_p + \mu_p) \quad \text{and} \quad \beta_p = \tfrac{1}{2}(\sigma_p - \mu_p).
$$

Since $(\Delta^m \sigma)_p = \sigma_{m,p} \geqslant D_{m,\,p}^{(k)} \geqslant \mid (\Delta^m \mu)_p \mid$ we have that the sequences $\alpha$ and $\beta$ are totally monotone, moreover, $\mu_p = \alpha_p - \beta_p$. $\triangle$

3.38 LEMMA. Let $\mu = \{\mu_k\}_0^\infty$ be a real sequence. If

$$
\sup_n \left\{ \sum_{k=0}^{n} \binom{n}{k} \mid (\Delta^{n-k}\mu)_k \mid \right\} \leqslant M \text{ for some } M > 0 \text{ then}
$$

$$
\lim_{n \to \infty} \binom{n}{k} (\Delta^{n-k}\mu)_k = 0 \text{ for } k = 1, 2, \ldots .
$$

*Proof.* By Lemma 3.37 $\mu$ is the difference of two totally monotone sequences. Thus, it is sufficient to assume that $\mu$ is totally monotone for our argument. Since $(\Delta^{n-k}\mu)_k = (\Delta^{n-k+1}\mu)_k + (\Delta^{n-k}\mu)_{k+1}$ we have that $\binom{n}{k}(\Delta^{n-k}\mu)_k = \binom{n}{k}(\Delta^{n-k+1}\mu)_k + \binom{n}{k}(\Delta^{n-k}\mu)_{k+1}$.
Therefore

$$(n+1)\binom{n}{k}(\Delta^{n-k}\mu)_k$$

$$= (n-k+1)\binom{n+1}{k}(\Delta^{n-k+1}\mu)_k + (k+1)\binom{n+1}{k+1}(\Delta^{n-k}\mu)_{k+1},$$

i.e.

$$(n+1)\left[\binom{n}{k}(\Delta^{n-k}\mu)_k - \binom{n+1}{k}(\Delta^{n-k+1}\mu)_k\right]$$

$$= (k+1)\binom{n+1}{k+1}(\Delta^{n-k}\mu)_{k+1} - k\binom{n+1}{k}(\Delta^{n-k+1}\mu)_k.$$

Now,

$$(n+1)\sum_{j=0}^{k}\left[\binom{n}{j}(\Delta^{n-j}\mu)_j - \binom{n+1}{j}(\Delta^{n-j+1}\mu)_j\right]$$

$$= \sum_{j=0}^{k}\left[(j+1)\binom{n+1}{j+1}(\Delta^{n-j}\mu)_{j+1} - j\binom{n+1}{j}(\Delta^{n-j+1}\mu)_j\right],$$

i.e.

$$(n+1)\left\{\sum_{j=0}^{k}\binom{n}{j}(\Delta^{n-j}\mu)_j - \sum_{j=0}^{k}\binom{n+1}{j}(\Delta^{n-j+1}\mu)_j\right\}$$

$$= (k+1)\binom{n+1}{k+1}(\Delta^{n-k}\mu)_{k+1} \geqslant 0.$$

So $\left\{\sum_{j=0}^{k}\binom{n}{j}(\Delta^{n-j}\mu)_j\right\}_{n=k}$ is a decreasing positive sequence. Thus, this sequence tends to a limit. Therefore,

$$\lim_{n\to\infty}\binom{n}{j}(\Delta^{n-k}\mu)_k = \lim_{n\to\infty}\left[\sum_{j=0}^{k}\binom{n}{j}(\Delta^{n-j}\mu)_j - \sum_{j=0}^{k-1}\binom{n}{j}(\Delta^{n-j}\mu)_j\right]$$

exists for each $k = 1, 2, \ldots$ (call the limit $t_k$). Now,

$$\sum_{j=0}^{k}\binom{n}{j}(\Delta^{n-j}\mu)_j - \sum_{j=0}^{k}\binom{n+1}{j}(\Delta^{n-j+1}\mu)_j = \frac{k+1}{n+1}\binom{n+1}{k+1}(\Delta^{n-k}\mu)_{k+1}$$

so, defining

$$p_n = \sum_{j=0}^{k}\binom{n}{j}(\Delta^{n-j}\mu)_j - \sum_{j=0}^{k}\binom{n+1}{j}(\Delta^{n-j+1}\mu)_j$$

we have that $p_n \sim \dfrac{k+1}{n+1}t_{k+1}$ $(n\to\infty)$. Since $\sum_{n=0}^{\infty}p_n$ converges it follows that $t_{k+1} = 0$ $(k = 0, 1, \ldots)$, i.e. $\lim_{n \to \infty} \binom{n}{k} (\Delta^{n-k}\mu)_k = 0$ for $k = 1, 2, \ldots$ . $\triangle$

3.39 THEOREM. *The Hausdorff matrix* $(H, \mu)$ *is regular if and only if*

(i) $\lim_{n \to \infty} (\Delta^n \mu)_0 = 0,$

(ii) $\mu_0 = 1,$ *and*

(iii) $\mu$ *is the difference of two totally monotone sequences.*

*Proof.* By Theorem 2.3 we have $(H, \mu) = (h_{n,k})$ is regular if and only if

$$\lim_{n \to \infty} h_{n,k} = 0 \text{ for each } k = 0, 1, \ldots \tag{3.7}$$

$$\lim_{n \to \infty} \sum_{k=0}^{\infty} h_{n,k} = 1, \tag{3.8}$$

and $$\sup_n \left\{ \sum_{k=0}^{\infty} |h_{n,k}| \right\} \leqslant M \text{ for some } M > 0. \tag{3.9}$$

Assuming (3.7), (3.8), and (3.9) we have that $\mu$ is the difference of two totally monotone sequences (by Lemma 3.37 and (3.9)), $\lim_{n \to \infty} (\Delta^n \mu)_0 = 0$ (by (3.7)), and $\mu_0 = 1$ (by the Exercise following Lemma 3.27). Therefore (i), (ii) and (iii) hold.

Now, assuming (i), (ii), and (iii) we have that (3.9) holds (by Lemma 3.37), (3.8) holds (by (ii) and the Exercise following Lemma 3.27), and (3.7) holds (by (i) and Lemmas 3.37 and 3.38). $\triangle$

We conclude this Chapter with an inclusion theorem for Hausdorff matrices.

3.40 THEOREM. *Let* $(H, \mu)$ *and* $(H, \mu')$ *be two Hausdorff matrices where* $\mu_n \neq 0$ *for* $n = 0, 1, \ldots$ $(\mu = \{\mu_n\}_0^\infty$ *and* $\mu' = \{\mu_n'\}_0^\infty)$. *If* $(H, \mu'/\mu)$ *is regular then* $(H, \mu) \subset (H, \mu')$.

*Proof.* Since $(H, \mu)$ and $(H, \mu')$ are lower triangular matrices and $(H, \mu)^{-1} = (H, 1/\mu)$ exists (by Corollary 3.32) we have that $(H, \mu) \subset (H, \mu')$ provided $(H, \mu') (H, \mu)^{-1}$ is regular, i.e. provided $(H, \mu'/\mu)$ is regular (by the proof of Lemma 3.31). $\triangle$

## CHAPTER 4

## Tauberian Theorems

In 1897 A. Tauber proved the theorem: If the real series $\sum_{n=0}^{\infty} a_n$ is Abel summable and if $a_n = o(1/n)$ then the series converges (see 'Ein Satz aus der Theorie der Unendlichen Reihen', *Monatsh. Math. Phys.* volume 8 (1897), pages 273-7). Theorems of this type in which ordinary convergence is deduced from the fact that one has some type of summability condition plus, perhaps, an additional condition (e.g. $a_n = o(1/n)$) are called Tauberian theorems. In this chapter we shall investigate some Tauberian theorems. In some cases a theorem might follow directly from other theorems which have been or are about to be proved, however, due to the importance of the methods of proof, we include independent proofs (see, for example, the fact that Theorem 4.11 follows from Theorem 4.13, however, the methods used to prove these two theorems are independent of each other and are individually important). Also, in this chapter, we assume that $a_n$ is real. Most of the theorems are true in the complex case as is seen, for example, by examining Theorem 4.16. If we let $a_n$ be complex, say $a_n = \alpha_n + i\beta_n$, and assume that $\sum_{n=0}^{\infty} a_n$ is $(C,1)$ convergent and $a_n = O(1/n)$ then we have that $\sum_{n=0}^{\infty} \alpha_n$ and $\sum_{n=0}^{\infty} \beta_n$ are each $(C,1)$ convergent, $a_n = O(1/n)$, and $\beta_n = O(1/n)$ and, therefore, by Theorem 4.15 we know that $\sum_{n=0}^{\infty} \alpha_n$ and $\sum_{n=0}^{\infty} \beta_n$ each converge, i.e. $\sum_{n=0}^{\infty} (\alpha_n + i\beta_n)$ converges.

## Elementary Tauberian Theorems based upon the Cesàro and Abel Methods

We begin by examining three theorems concerning sequences of real numbers. These will allow us to examine Tauberian theorems associated with the Cesàro and Abel methods of summability (refer to Chapters 1 and 3 for these methods).

4.1  THEOREM. *If* $\{c_n\}_0^\infty$ *is a sequence of positive numbers increasing to* $+\infty$ *and if* $\{a_n\}_0^\infty$ *is any sequence of real numbers then*

$$\lim \inf_{n\to\infty} \frac{a_{n+1}-a_n}{c_{n+1}-c_n} \leqslant \lim \inf_{n\to\infty} \frac{a_n}{c_n} \leqslant \lim \sup_{n\to\infty} \frac{a_n}{c_n}$$

$$\leqslant \lim \sup_{n\to\infty} \frac{a_{n+1}-a_n}{c_{n+1}-c_n} \ .$$

We omit the proof (see Example 1.17 and Exercise (6) following Example 1.21).

4.2  COROLLARY. *If* $\{a_n\}_0^\infty$ *is any sequence of real numbers then*

$$\lim \inf_{n\to\infty} (a_{n+1}-a_n) \leqslant \lim \inf_{n\to\infty} \frac{a_n}{n} \leqslant \lim \sup_{n\to\infty} \frac{a_n}{n}$$

$$\leqslant \lim \sup_{n\to\infty} (a_{n+1}-a_n).$$

The proof follows directly from Theorem 4.1 by letting $c_n = n$.

4.3  COROLLARY. *If* $\{S_n\}_0^\infty$ *is any sequence of real numbers then*

$$\lim \inf_{n\to\infty} S_n \leqslant \lim \inf_{n\to\infty} \frac{S_0 + \ldots + S_n}{n+1}$$

$$\leqslant \lim \sup_{n\to\infty} \frac{S_0 + \ldots + S_n}{n+1} \leqslant \lim \sup_{n\to\infty} S_n. \quad (4.1)$$

To prove this corollary let $a_n = \sum_{k=0}^{n} S_k$ and $c_n = n+1$ in Theorem 4.1.

Before continuing recall, from Chapter 3 (Definition 3.6 and following material), the notation

$$A_n^\alpha = \frac{(\alpha+1) \ldots (\alpha+n)}{n!}$$

and $S_n^k = \sum_{j=0}^{n} S_j^{k-1}$ where $k$ is a positive integer and $S_n^0 = S_n = \sum_{j=0}^{n} a_j$.

Now, if $\lim_{n\to\infty} S_n^k/A_n^k = y$, then the series $\sum_{j=0}^{\infty} a_j$ is $(C,k)$-summable to $y$. If we introduce the gamma function we have that

$$A_n^\alpha = \frac{\Gamma(\alpha+n+1)}{\Gamma(\alpha+1)\Gamma(n+1)}$$

and, hence, the series $\sum_{j=0}^{\infty} a_j$ is $(C,k)$-summable to $y$ ($k$ a positive integer) provided

$$\lim_{n\to\infty} \frac{S_n^k}{\Gamma(k+n+1)/\Gamma(k+1)\Gamma(n+1)} = y.$$

**4.4 THEOREM.** *Let $k$ be a positive integer. If $\sum_{n=0}^{\infty} a_n$ is $(C,k)$-summable then $a_n = o(n^k)$ and $S_n = o(n^k)$.*

*Proof.* By hypothesis there exists $S$ such that

$$\lim_{n\to\infty} \frac{S_n^k}{\Gamma(k+n+1)/\Gamma(n+1)\Gamma(k+1)} = S_n.$$

Hence,

$$S_n^k = \{\frac{n^k}{\Gamma(k+1)} (1 + o(1))\} (S + o(1)) = S \frac{n^k}{\Gamma(k+1)} + o(n^k) \quad (4.2)$$

and so

$$S_n^{k-1} = S_n^k - S_{n-1}^k = o(n^k).$$

Similarly $S^{k-j} = o(n^k)$ for $j = 2, 3, \ldots, k$. Therefore $a_n = S_n - S_{n-1} = o(n^k)$. $\triangle$

**4.5 THEOREM.** *If the series $\sum_{n=0}^{\infty} a_n$ is $(C,1)$-summable to $S$ and $a_n \geqslant 0$ for $n$ sufficiently large then $\sum_{n=0}^{\infty} a_n$ converges to $S$.*

*Proof.* Let $S_n = \sum_{j=0}^{n} a_j$. There exists $n_0$ such that if $n \geqslant n_0$ then $S_{n+1} \geqslant S_n$. If $S_n = O(1)$ then $\{S_n\}_0^\infty$ converges (say to $L$) and (4.1) implies that $L = S$.

If $\{S_n\}_0^\infty$ is unbounded then $\lim_{n\to\infty} S_n = +\infty$ and (4.1) implies that $\lim_{n\to\infty} S_n^1/(n+1) = +\infty$ which contradicts the hypothesis. $\triangle$

**4.6 LEMMA.** *If $\sum_{n=0}^{\infty} a_n x^n$ converges for $|x| < 1$ and $\sum_{n=0}^{\infty} a_n = +\infty$ then $\lim_{x\to 1-} \sum_{n=0}^{\infty} a_n x^n = +\infty$.*

*Proof.* Let $f(x) = \sum_{n=0}^{\infty} a_n x^n$ for $|x| < 1$. We have, for $|x| < 1$, that

$$f(x) = f(x)(1-x)\sum_{n=0}^{\infty} x^n. \quad \text{So}$$

$$f(x) = (1-x)\sum_{n=0}^{\infty} S_n x^n \quad \text{where} \quad S_n = \sum_{j=0}^{n} a_j. \quad (4.3)$$

Since $\{S_n\}_0^{\infty} \to +\infty$ we have, given $H > 0$, that there exists $n_1 = n_1(H)$ such that $S_n > H$ for $n \geqslant n_1$. So, for $0 < x < 1$,

$$(1-x)\sum_{n=n_1+1}^{\infty} S_n x^n > H(1-x)\left(\sum_{n=n_1+1}^{\infty} x^n\right) = Hx^{n_1+1}.$$

Now, there exists $x_1$ such that if $0 < x_1 < x < 1$ then $Hx^{n_1+1} > \frac{1}{2}H$. Furthermore, there exists $x_2$ such that if $0 < x_2 < x < 1$ then

$$(1-x)\left|\sum_{n=0}^{n_1} S_n x^n\right| < (1-x)\sum_{n=0}^{n_1}|S_n| < \frac{1}{4}H.$$

Hence, for $0 < \max\{x_1, x_2\} < x < 1$,

$$f(x) > (1-x)\sum_{n=n_1+1}^{\infty} S_n x^n - (1-x)\left|\sum_{n=0}^{n_1} S_n x^n\right| > \frac{1}{4}H.$$

Since $H$ is arbitrary we have that

$$\lim_{x \to 1-}\sum_{n=0}^{\infty} a_n x^n = \lim_{x \to 1-} f(x) = +\infty. \quad \triangle$$

**4.7 THEOREM.** *If the series* $\sum_{n=0}^{\infty} a_n$ *is Abel summable to S and* $a_n \geqslant 0$ *for n sufficiently large then* $\sum_{n=0}^{\infty} a_n$ *converges to S.*

*Proof.* There exists $n_0$ such that if $n > n_0$ then $a_n > 0$. Thus the sequence $\{S_n\}_{n_0+1}^{\infty}$ is an increasing sequence. If $S_n = O(1)$ then there exists $L$ such that $\sum_{n=0}^{\infty} a_n = L$. By Theorem 1.7 we have that

$$\lim_{x \to 1-}\sum_{n=0}^{\infty} a_n x^n = \sum_{n=0}^{\infty} a_n$$

and, hence, $S = L$.

If $S_n \neq O(1)$ then $\lim_{n \to \infty} S_n = +\infty$ and, by Lemma 4.6, the series $\sum_{n=0}^{\infty} a_n$ is not Abel summable (which contradicts the hypothesis). $\quad \triangle$

We will now prove that a series which is $(C,k)$-summable is also Abel summable. This will give us an alternate proof to Theorem 4.5.

4.8 THEOREM. *Let $k$ be a positive integer. If $\sum\limits_{n=0}^{\infty} a_n$ is $(C,k)$-summable to $S$ then $\sum\limits_{n=0}^{\infty} a_n$ is Abel summable to S.*

*Proof.* Since $\sum\limits_{n=0}^{\infty} a_n$ is $(C,k)$-summable we have, by Theorem 4.4, that $a_n = o(n^k)$. Thus the series $\sum\limits_{n=0}^{\infty} a_n x^n$ converges for $|x| < 1$ (say to $f(x)$). As in (4.3) we have that

$$f(x) = \sum_{n=0}^{\infty} a_n x^n = (1-x) \sum_{n=0}^{\infty} S_n x^n$$

$$= (1-x)^2 \sum_{n=0}^{\infty} S_n^1 x^n = \ldots = (1-x)^{k+1} \sum_{n=0}^{\infty} S_n^k x^n. \qquad (4.4)$$

Since $n^k/k! - \binom{n}{k} = o(n^k)$ we have, from (4.2), that

$$S_n^k = S\binom{n}{k} + o(n^k) = S\binom{n}{k} + o(\binom{n}{k}).$$

Hence, given $\varepsilon > 0$, there exists $N = N(\varepsilon) \geqslant k$ such that

$$\left| S_n^k - S\binom{n}{k} \right| < \varepsilon \binom{n}{k} \quad \text{provided } n \geqslant N.$$

If $0 < x < 1$ then

$$\left| \sum_{n=0}^{\infty} S_n^k x^n - S \sum_{n=k}^{\infty} \binom{n}{k} x^n \right|$$

$$\leqslant \sum_{n=k}^{\infty} \left| S_n^k - S\binom{n}{k} \right| x^n + \sum_{n=0}^{k-1} \left| S_n^k \right| x^n$$

$$\leqslant \varepsilon \sum_{n=N+1}^{\infty} \binom{n}{k} x^n + \sum_{n=0}^{k-1} \left| S_n^k \right| + \sum_{n=k}^{N} \left| S_n^k - S\binom{n}{k} \right|$$

$$\leqslant \varepsilon \sum_{n=k}^{\infty} \binom{n}{k} x^n + \sum_{n=0}^{k-1} \left| S_n^k \right| + \sum_{n=k}^{N} \left| S_n^k - S\binom{n}{k} \right|.$$

Now $\sum\limits_{n=k}^{\infty} \binom{n}{k} x^n = x^k (1-x)^{-k-1}$ hence

$$\limsup_{x \to 1-} \frac{\left| \sum\limits_{n=0}^{\infty} S_n^k x^n - S \sum\limits_{n=k}^{\infty} \binom{n}{k} x^n \right|}{x^k (1-x)^{-k-1}} \leqslant \varepsilon.$$

Therefore

$$\lim_{x \to 1-} \frac{\sum_{n=0}^{\infty} S_n^k x^n - S \sum_{n=k}^{\infty} \binom{n}{k} x^n}{x^k (1-x)^{-k-1}} = 0.$$

Using (4.4) we have that

$$\lim_{x \to 1-} \frac{f(x) (1-x)^{-k-1}}{x^k (1-x)^{-k-1}} = S,$$

i.e.    $\lim_{x \to 1-} f(x) = \lim_{x \to 1-} \sum_{n=0}^{\infty} a_n x^n = S.$   $\triangle$

An immediate consequence of Theorem 4.8 and Theorem 3.11 is

**4.9 COROLLARY.** *Let $\alpha > -1$. If $\sum_{n=0}^{\infty} a_n$ is $(C, \alpha)$-summable to $S$ then $\sum_{n=0}^{\infty} a_n$ is Abel summable to $S$.*

**4.10 COROLLARY.** *Let $\alpha > -1$. If $\sum_{n=0}^{\infty} a_n$ is $(C, \alpha)$-summable to $S$ and $a_n \geqslant 0$ for sufficiently large $n$ then $\sum_{n=0}^{\infty} a_n$ converges to $S$.*

*Proof.* The series is Abel summable to $S$ (by Corollary 4.9) hence it converges to $S$ (by Theorem 4.7).   $\triangle$

**4.11 THEOREM.** *Let $T_n = \sum_{j=1}^{n} j a_j$. If $\sum_{n=0}^{\infty} a_n$ is $(C, 1)$-summable to $S$ then $\sum_{n=0}^{\infty} a_n$ converges to $S$ if and only if $T_n = o(n)$.*

*Proof.* We have that

$$(n+1) S_n - S_n^1 = (n+1) \sum_{j=0}^{n} a_j - \sum_{j=0}^{n} S_j = T_n.$$

Hence

$$S_n - \frac{S_n^1}{n+1} = \frac{T_n}{n+1}. \tag{4.5}$$

If $\sum_{n=0}^{\infty} a_n$ converges then, by (4.1),

$$\lim_{n \to \infty} \frac{S_n^1}{n+1} = \lim_{n \to \infty} S_n$$

thus $T_n = o(n)$.

If $T_n = o(n)$ then, by (4.5) and our hypothesis, we have that

$$S_n = \frac{S_n^1}{n+1} + o(1).$$

So $\sum\limits_{n=0}^{\infty} a_n$ converges to S. $\triangle$

4.12 COROLLARY. If $\sum\limits_{n=0}^{\infty} a_n$ is $(C,1)$-summable to S and $a_n = o(1/n)$ then $\sum\limits_{n=0}^{\infty} a_n$ converges to S.

Proof. Since $na_n = o(1)$ we have, by (4.1), that $T_n = o(n)$. So, $\sum\limits_{n=0}^{\infty} a_n$ converges to S by Theorem 4.11. $\triangle$

4.13 THEOREM. If $\sum\limits_{n=0}^{\infty} a_n$ is Abel summable to S then $\sum\limits_{n=0}^{\infty} a_n$ to S if and only if $T_n = o(n)$.

Proof. (a) First assume that $na_n = o(1)$. If $0 < x < 1$ then

$$f(x) - S_m = \sum_{n=0}^{\infty} a_n x^n - \sum_{n=0}^{m} a_n$$

$$= - \sum_{n=0}^{m} a_n (1-x^n) + \sum_{n=m+1}^{\infty} a_n x^n.$$

Now $1-x^n = (1-x) \sum\limits_{j=0}^{n-1} x^j < n(1-x)$. If $\varepsilon_m = \max\limits_{k \geqslant m} |ka_k|$ then

$$|f(x) - S_m| \leqslant \sum_{n=1}^{m} (1-x) n |a_n| + \left| \sum_{n=m+1}^{\infty} \frac{na_n x^n}{n} \right|$$

$$\leqslant (1-x) \sum_{n=1}^{m} n |a_n| + \varepsilon_{m+1} \sum_{n=m+1}^{\infty} \frac{x^n}{n}$$

$$< (1-x) \sum_{n=1}^{m} n |a_n| + \frac{\varepsilon_{m+1}}{m+1} \frac{1}{1-x}.$$

Since $na_n = o(1)$ we have, by (4.1), that $\sum\limits_{n=0}^{m} na_n = o(m)$. Also, $\{\varepsilon_m\}_0^{\infty}$ decreases to 0. Therefore,

$$\left| f(1-\tfrac{1}{m}) - S_m \right| \leqslant \frac{1}{m} \sum_{n=0}^{m} n |a_n| + \varepsilon_{m+1} = o(1),$$

i.e. $\lim\limits_{m \to \infty} S_m = S$.

(b)  Now we prove the theorem when $T_n = o(n)$.  We have that $T_0 = 0$ and $(T_n - T_{n-1})/n = a_n$ for $n \geqslant 1$.  Hence, for $0 < x < 1$,

$$f(x) = a_0 + \sum_{n=1}^{\infty} \frac{T_n}{n} x^n - \sum_{n=1}^{\infty} \frac{T_{n-1}}{n} x^n$$

$$= a_0 + \sum_{n=1}^{\infty} \frac{T_n}{n} x^n - \sum_{n=1}^{\infty} \frac{T_n}{n+1} x^{n+1}$$

$$= a_0 + \sum_{n=1}^{\infty} \frac{T_n}{n} x^n - \sum_{n=1}^{\infty} \left( \frac{T_n}{n} - \frac{T_n}{n(n+1)} \right) x^{n+1}$$

$$= a_0 + (1-x) \sum_{n=1}^{\infty} \frac{T_n}{n} x^n + \sum_{n=1}^{\infty} \frac{T_n}{n(n+1)} x^{n+1} .$$

Let $\max_{n \geqslant m} |T_n|/n = \delta_m$.   Thus $\{\delta_m\}_0^{\infty}$ decreases to 0 and

$$\limsup_{x \to 1-} \left| (1-x) \sum_{n=1}^{\infty} \frac{T_n}{n} x^n \right|$$

$$\leqslant \limsup_{x \to 1-} (1-x) \sum_{n=1}^{m} \frac{|T_n|}{n} x^n + \limsup_{x \to 1-} (1-x) \frac{\delta_{m+1}}{1-x} = \delta_{m+1} .$$

Hence

$$\lim_{x \to 1-} (1-x) \sum_{n=1}^{\infty} \frac{T_n}{n} x^n = 0$$

and

$$\lim_{x \to 1-} f(x) = a_0 + \sum_{n=1}^{\infty} \frac{T_n}{n(n+1)} x^{n+1} .$$

So, the series

$$\sum_{n=1}^{\infty} \frac{T_n}{n(n+1)} \tag{4.6}$$

is Abel summable to $S - a_0$ and $T_n/[n(n+1)] = o(1/n)$.  Consequently, by (a) the series (4.6) converges to $S - a_0$.   But the sum of the series in (4.6) is

$$\lim_{n \to \infty} \sum_{k=1}^{n} \frac{T_k}{k(k+1)} = \lim_{n \to \infty} \sum_{k=1}^{n} \left( \frac{T_k}{k} - \frac{T_k}{k+1} \right)$$

$$= \lim_{n \to \infty} \left( \sum_{k=1}^{n} \frac{T_k}{k} - \sum_{k=2}^{n+1} \frac{T_{k-1}}{k} \right)$$

$$= \lim_{n \to \infty} \left( -\frac{T_n}{n+1} + \sum_{k=1}^{n} a_k \right) = \lim_{n \to \infty} \sum_{k=1}^{n} a_k.$$

Therefore, $\sum_{k=0}^{\infty} a_k$ converges to S.

To complete the proof of the theorem we note that if $\sum_{n=0}^{\infty} a_n$ converges then $T_n = o(n)$ (by Theorem 4.11).  $\triangle$

The proofs of the following two corollaries are immediate.

4.14  COROLLARY.  *Let* $\cdot \alpha > -1$. *If* $\sum_{n=0}^{\infty} a_n$ *is* $(C, \alpha)$-*summable to* S *and* $T_n = o(n)$ *then* $\sum_{n=0}^{\infty} a_n$ *converges to* S.

4.15  COROLLARY.  *If* $\sum_{n=0}^{\infty}$ *is Abel summable to* S *and* $a_n = o(1/n)$ *then* $\sum_{n=0}^{\infty} a_n$ *converges to* S.

### Tauberian Theorems continued

In the next group of theorems we relax the condition on $a_n$; for example, Theorem 4.16 which is due to G.H. Hardy who replaced the Tauberian condition $a_n = o(1/n)$ (of Corollary 4.12) by $a_n = O(1/n)$. This result has been extended by E. Landau who assumed the one-sided Tauberian condition $na_n \geqslant -c$ ($c > 0$) (Theorem 4.18).  For Theorem 4.16 we give Szasz's proof (see O. Szasz, *Introduction to the Theory of Divergent Series*, chapter 4).

4.16  THEOREM.  *If* $\sum_{n=0}^{\infty} a_n$ *is* $(C, 1)$-*summable to* S *and* $a_n = O(1/n)$ *then* $\sum_{n=0}^{\infty} a_n$ *converges to* S.

*Proof.*  Let $T_n = \sum_{k=1}^{n} ka_k$.  We have that

$$\sum_{n=1}^{N} \frac{T_n}{n(n+1)} = \sum_{n=1}^{N} \left( \frac{1}{n} - \frac{1}{n+1} \right) T_n$$

$$= \sum_{n=1}^{N} \frac{T_n - T_{n-1}}{n} - \frac{T_N}{N+1}$$

$$= \sum_{n=1}^{N} a_n - \frac{T_N}{N+1} = S_N - a_0 - (S_N - \frac{S_N^1}{N+1})$$

$$= S - a_0 + o(1).$$

Thus, the series $\sum_{n=1}^{\infty} \frac{T_n}{n(n+1)}$ converges.

We now show that $T_n = o(n)$. Suppose that $T_n \neq o(n)$. Then there exists $A > 0$ such that either

(i) $T_n > An$ for an infinite number of values of $n$; or

(ii) $T_n < -An$ for an infinite number of values of $n$.

Suppose case (i) holds, i.e. there exist

$$0 < n_1 < n_2 < \ldots < n_k < \ldots \quad \text{where } T_{n_k} > An_k.$$

We have that there exists $c > 0$ such that $n \mid a_n \mid \leqslant c$ hence $(n+1)a_{n+1} > -c$ for all $n \geqslant 1$. Hence

$$T_{n+1} = T_n + (n+1)a_{n+1} > T_n - c$$

and, in general, $T_{n+\nu} > T_n - \nu c > An - \nu c \geqslant \frac{1}{2} An$ where $0 \leqslant \nu \leqslant \frac{1}{2} (An/c)$ and $n = n_k$.

Writing $n_k = M$ we have that

$$\sum_{n=M}^{M+[\frac{1}{2}(AM/c)]} \frac{T_n}{n(n+1)} > \frac{A}{2} \sum_{n=M}^{M+[\frac{1}{2}(AM/c)]} \frac{1}{n+1}$$

$$> \frac{A}{2} \frac{1+[\frac{1}{2}(AM/c)]}{M+[\frac{1}{2}(AM/c)]+1} > \frac{A}{2} \frac{\frac{1}{2}(AM/c)}{M+\frac{1}{2}(AM/c)} = \frac{A^2}{4c+2A}.$$

This contradicts the fact that $\sum_{n=1}^{\infty} T_n/[n(n+1)]$ converges.

Case (ii) is handled similarly.

Therefore $T_n = o(n)$ and, by Theorem 4.11, $\sum_{n=0}^{\infty} a_n$ converges to $S$. $\triangle$

*Exercise.* Prove that $\sum_{n=1}^{\infty} a_k$ is (C,1)-summable if and only if $\sum_{n=1}^{\infty} T_n/[n(n+1)]$ is convergent.

Before proving Landau's theorem we require

4.17 DEFINITION. *Let* $x = \{x_n\}_0^\infty$ *be a bounded sequence. The oscillation function,* $\omega_n(x)$, *of the sequence* $\{x_n\}_0^\infty$ *is*

$$\omega_n(x) = \sup_{k,\ell \geqslant n} |x_k - x_\ell|.$$

It is readily seen that $\omega_n(x)$ is a non-increasing function of $n$ and, for $x_n$ real,

$$\lim_{n \to \infty} \omega_n(x) = \limsup_{n \to \infty} x_n - \liminf_{n \to \infty} x_n.$$

4.18 THEOREM. *If* $\sum\limits_{n=0}^{\infty} a_n$ *is* (C,1)-*summable to* $S$ *and* $na_n \geqslant -c$ (*some* $c > 0$) *then* $\sum\limits_{n=0}^{\infty} a_n$ *converges to* $S$.

*Proof.* For $S_n = \sum\limits_{k=0}^{n} a_k$ we have that

$$S_n - \frac{S_{n+k}^1}{n+k+1} - \frac{n}{k+1}\left\{\frac{S_{n+k}^1}{n+k+1} - \frac{S_{n-1}^1}{n}\right\}$$

$$= S_n - \frac{n+k+1}{k+1}\frac{1}{n+k+1}\sum_{j=0}^{n+k} S_j + \frac{1}{k+1}\sum_{j=0}^{n-1} S_j$$

$$= S_n - \frac{1}{k+1}\sum_{j=n}^{n+k} S_j$$

$$= -\frac{1}{k+1}\{ka_{n+1} + (k-1)a_{n+2} + \ldots + a_{n+k}\}.$$

Writing $\sigma_n = S_n^1/(n+1)$ and $\omega_n = \omega_n(\sigma)$ where $\sigma = \{\sigma_n\}_0^\infty$ we have that

$$S_n - \sigma_{n+k} = \frac{n}{k+1}(\sigma_{n+k} - \sigma_{n-1}) - \frac{1}{k+1}\sum_{j=1}^{k}(k+1-j)a_{n+j}. \qquad (4.7)$$

Therefore

$$S_n \leqslant \sigma_{n+k} + \frac{n}{k+1}\omega_{n-1} + \frac{c}{k+1}\left\{\frac{k}{n+1} + \frac{k-1}{n+2} + \ldots + \frac{1}{n+k}\right\}$$

$$\leqslant \sigma_{n+k} + \frac{n}{k+1}\omega_{n-1} + \frac{c}{2}\frac{k(k+1)}{(k+1)(n+1)}.$$

Choose $k = k(n) = [n\sqrt{(\omega_{n-1})}]$. Since $\{\omega_n\}_0^\infty$ decreases to 0 we have that

$$\lim \sup_{n \to \infty} S_n \leqslant \lim_{n \to \infty} \sigma_n = S.$$

We now estimate $\lim \inf_{n \to \infty} S_n$. If $k < n-1$ then one can verify that

$$S_n - \sigma_{n-k-1} = \frac{n+1}{k+1} (\sigma_n - \sigma_{n-k-1}) + \frac{1}{k+1} \sum_{j=n-k+1}^{n} (j-n+k) \, a_j$$

thus

$$S_n \geqslant \sigma_{n-k-1} - \frac{n+1}{k+1} \omega_{n-k-1} - \frac{c}{k+1} \frac{k(k+1)}{2(n-k+1)} .$$

Given $\varepsilon > 0$ where $\varepsilon < \frac{1}{2}$ and letting $k = k(n) = [\varepsilon n]$ we have that $\lim_{n \to \infty} \frac{n+1}{k+1} \omega_{n-k-1} = 0$ and hence

$$\lim \inf_{n \to \infty} S_n \geqslant S - \varepsilon c.$$

This gives us the fact that

$$\lim \inf_{n \to \infty} S_n \geqslant S.$$

Combining this with (4.8) we have that $\lim_{n \to \infty} S_n = S.$ $\triangle$

We see that $\sum_{n=0}^{\infty} a_n$ converges to $S$ provided $\sum_{n=0}^{\infty} a_n$ is $(C,1)$-summable to $S$ and $na_n \leqslant c$ for some $c > 0$ (using Theorem 4.18 and considering the series $\sum_{n=0}^{\infty} (-a_n)$).

Also, using (4.7), we have an alternate proof to Theorem 4.16. Assuming the hypotheses of Theorem 4.16 we have that $\{n|a_n|\}_1^{\infty}$ is bounded. Let

$$\delta_n = \sup_{k \geqslant n} |a_k| .$$

Thus $\{\delta_n\}_0^{\infty}$ decreases to 0 and $n\delta_n = O(1)$. From (4.7) we have that

$$|S_n - \sigma_{n+k}| \leqslant \frac{n}{k+1} \omega_{n-1} + \frac{k}{2} \delta_{n+1} .$$

If

$$k = k(n) = \left[ \frac{2n \, \omega_{n-1}}{\delta_{n+1}} \right]^{1/2}$$

then

$$|S_n - \sigma_{n+k}| < (2n \, \omega_{n-1} \, \delta_{n+1})^{1/2}.$$

Therefore $\lim_{n \to \infty} S_n = S$ since $\lim_{n \to \infty} (2n\, \omega_{n-1}\, \delta_{n+1})^{1/2} = 0$.

We now prove the famous Hardy–Littlewood theorem (Theorem 4.22) which states that the series $\overset{\infty}{\underset{n=0}{\Sigma}}\, a_n$ converges provided $\overset{\infty}{\underset{n=0}{\Sigma}}\, a_n$ is Abel summable and $na_n \geqslant -c$ for some $c > 0$. The proof depends upon an ingenious argument due to Karamata in 1930 (see J. Karamata, 'Uber die Hardy-Littlewoodschen Umkehrungen des Abelschen Stetigkeitsatzes', *Math. Z.* volume 32, pages 319-20). We develop this argument in the following preparatory theorems (one of which uses the Weierstrass Approximation theorem; see Corollary 5.40). For the method used by Hardy and Littlewood refer to G. H. Hardy, *Divergent Series,* chapter 7.

In the following we require a function having, at most, a discontinuity of the first kind. A function $F$, has a discontinuity of the first kind at $x_0$ if $F$ is discontinuous at $x_0$, $F(x_0)$ is defined, and the left-hand limit $F(x_0-)$ and the right-hand limit $F(x_0+)$ exist.

4.19 LEMMA. *Let $F(x)$ be continuous on $[0,1]$ except for, at most, one discontinuity of the first kind on $(0,1)$. If $\varepsilon > 0$ is given then there exist two polynomials $p(x)$ and $P(x)$ such that*

$$p(x) \leqslant F(x) \leqslant P(x) \quad \text{for} \quad x \in [0,1] \quad\quad (4.9)$$

*and*

$$\int_0^1 \{F(x) - p(x)\}\, dx \leqslant \varepsilon \quad \text{and} \quad \int_0^1 \{P(x) - F(x)\}\, dx \leqslant \varepsilon. \quad (4.10)$$

*Proof.* First, assume that $F(x)$ is continuous. Thus the functions $F(x) \pm \frac{1}{2}\varepsilon$ are continuous and, by the Weierstrass Approximation theorem, there exist polynomials $p(x)$ and $P(x)$ such that

$$\left| p(x) - (F(x) - \tfrac{1}{2}\varepsilon) \right| \leqslant \tfrac{1}{2}\varepsilon \quad \text{and} \quad \left| P(x) - (F(x) + \tfrac{1}{2}\varepsilon) \right| \leqslant \tfrac{1}{2}\varepsilon$$

for $0 \leqslant x \leqslant 1$. Therefore, $p(x)$ and $P(x)$ satisfy (4.9) and (4.10).

Now, assume that $F(x)$ has a discontinuity of the first kind at $x_0$ (where $0 < x_0 < 1$). Let $\varepsilon > 0$ be given. Choose $\delta_1 > 0$ such that

$$\max_{x \in I_1} |F(x) - F(x_0-)| < \tfrac{1}{8}\varepsilon \quad \text{and} \quad \max_{x \in I_2} |F(x) - F(x_0+)| < \tfrac{1}{8}\varepsilon$$

where $I_1 = [x_0 - \delta_1,\, x_0[ \cap [0,1]$ and $I_2 = ]x_0,\, x_0 + \delta_1] \cap [0,1]$. Let

$$D = \max\{\max_{x \in I_1 \cup I_2} F(x),\, F(x_0)\} \quad \text{and}$$

$$d = \min\{\min_{x \in I_1 \cup I_2} F(x),\, F(x_0)\}\,.$$

Choose $0 < \delta < \delta_1$ such that $2\delta(|D| + |d|) < \tfrac{1}{4}\varepsilon$ and let $\ell_1(x)$ be the linear function representing the line which passes through the points $(x_0-\delta, F(x_0-\delta) + \frac{1}{4}\varepsilon)$ and $(x_0, D + \frac{1}{4}\varepsilon)$ and $\ell_2(x)$ be the linear function representing the line which passes through the points $(x_0, D + \frac{1}{4}\varepsilon)$ and $(x_0+\delta, F(x_0+\delta) + \frac{1}{4}\varepsilon)$. Now, define the function $F_1(x)$ by

$$
F_1(x) = \begin{cases}
F(x) + \frac{1}{4}\varepsilon & \text{if } 0 \leqslant x \leqslant x_0 - \delta \text{ or } x_0 + \delta \leqslant x \leqslant 1 \\[2mm]
\ell_1(x) & \text{if } x_0 - \delta < x \leqslant x_0 \\[2mm]
\ell_2(x) & \text{if } x_0 < x < x_0 + \delta.
\end{cases}
$$

The function $F_1(x)$ is uniquely defined and continuous on $[0,1]$, moreover, $F_1(x) \geqslant F(x) + \frac{1}{4}\varepsilon$ on $[0,1]$. Now,

$$
\int_0^1 \{F_1(x) - F(x)\}\, dx \leqslant \frac{1}{4}\varepsilon + \int_{x_0-\delta}^{x_0+\delta} \{F_1(x) - F(x)\}\, dx
$$

$$
\leqslant \frac{1}{4}\varepsilon + [2\delta\,(|D| + |d|) + \frac{1}{4}\varepsilon] < \frac{3}{4}\varepsilon.
$$

Choose a polynomial $P(x)$ such that $P(x) \geqslant F_1(x)$ on $[0,1]$ and $\int_0^1 \{P(x) - F_1(x)\}\, dx \leqslant \frac{1}{4}\varepsilon$. Thus $P(x) > F(x)$ on $[0,1]$ and

$$
\int_0^1 \{P(x) - F(x)\}\, dx = \int_0^1 \{P(x) - F_1(x)\}\, dx + \int_0^1 \{F_1(x) - F(x)\}\, dx
$$

$$
< \frac{1}{4}\varepsilon + \frac{3}{4}\varepsilon = \varepsilon.
$$

Therefore, $P(x)$ satisfies condition (4.10).

We can construct a polynomial $p(x)$ satisfying condition (4.9) in a similar manner.  $\triangle$

**4.20 THEOREM.** *If the series* $\sum\limits_{n=0}^{\infty} a_n$ *is Abel summable to S and* $S_n = \sum\limits_{j=0}^{n} a_j \geqslant -c$ *(for some $c > 0$ and all $n = 0, 1, \ldots$ ) then the series* $\sum\limits_{n=0}^{\infty} a_n$ *is $(C,1)$-summable to S.*

*Proof.* We are given that

$$
\lim_{t\to 1-} \sum_{n=0}^{\infty} a_n t^n = S,
$$

i.e.

$$
\lim_{t\to 1-} (1-t) \sum_{n=0}^{\infty} S_n t^n = S. \tag{4.11}
$$

Thus

$$\lim_{t\to 1-} (1-t) \sum_{n=0}^{\infty} (S_n + c) t^n = S + c$$

where $S + c \geqslant 0$.   We must prove

$$\lim_{n\to\infty} \frac{1}{n+1} \sum_{\nu=0}^{n} S_\nu = S, \tag{4.12}$$

i.e.

$$\lim_{n\to\infty} \frac{1}{n+1} \sum_{\nu=0}^{n} (S_\nu + c) = S + c.$$

So, we may assume that $S_n \geqslant 0$ (otherwise, we would work with the sequence $\{\omega_n\}^\infty$ where $\omega_n = S_n + c$).

Now,   $\lim_{x\to 1-} (1-x) \sum_{n=0}^{\infty} S_n x^n (x^n)^k$

$$= \lim_{x\to 1-} (1-x) \sum_{n=0}^{\infty} S_n x^{n+kn}$$

$$= \lim_{x\to 1-} \left\{ \frac{1-x}{1-x^{k+1}} (1-x^{k+1}) \sum_{n=0}^{\infty} S_n (x^{k+1})^n \right\}$$

$$= \frac{1}{k+1} S = S \int_0^1 x^k \, dx$$

thus

$$\lim_{x\to 1-} (1-x) \sum_{n=0}^{\infty} S_n x^n P(x^n) = S \int_0^1 P(x) \, dx \tag{4.13}$$

where $P(x)$ is any polynomial.

Let $\varepsilon > 0$ be given and $F(x)$ be any function described in Lemma 4.19.  By Lemma 4.19 choose a polynomial $P(x)$ such that $P(x) \geqslant F(x)$ on $[0,1]$ and $\int_0^1 (P(x) - F(x)) \, dx \leqslant \varepsilon$.  Since $S_n \geqslant 0$ we have that

$$\limsup_{x\to 1-} (1-x) \sum_{n=0}^{\infty} S_n x^n F(x^n) \leqslant \lim_{x\to 1-} (1-x) \sum_{n=0}^{\infty} S_n x^n P(x^n)$$

$$= S \int_0^1 P(x) \, dx \leqslant S \left( \int_0^1 F(x) \, dx + \varepsilon \right).$$

Therefore, $\limsup_{x\to 1-} (1-x) \sum_{n=0}^{\infty} S_n x^n F(x^n) \leqslant S \int_0^1 F(x) \, dx$.

Similarly, using $p(x)$ and Lemma 4.19, we have that

$$\lim \inf_{x \to 1-} (1-x) \sum_{n=0}^{\infty} S_n x^n F(x^n) \geqslant S \int_0^1 F(x) \, dx.$$

So, we can replace $P(x)$ by $F(x)$ in (4.13), i.e.

$$\lim_{x \to 1-} (1-x) \sum_{n=0}^{\infty} S_n x^n F(x^n) = S \int_0^1 F(x) \, dx.$$

Define

$$F(x) = \begin{cases} 0 & \text{if } 0 \leqslant x < e^{-1} \\[2mm] \dfrac{1}{x} & \text{if } e^{-1} \leqslant x \leqslant 1. \end{cases}$$

Thus

$$\int_0^1 F(x) \, dx = 1 \quad \text{and} \quad F(x^n) = 0 \quad \text{for } n > (\log \tfrac{1}{x})^{-1}.$$

Let $x = e^{-1/N}$.    Thus

$$\lim_{N \to \infty} (1-x) \sum_{n=0}^{\infty} S_n x^n F(x^n) = \lim_{N \to \infty} (1-e^{-1/N}) \sum_{n=0}^{N} S_n = S.$$

Since $\lim_{N \to \infty} (N+1)(1-e^{-1/N}) = 1$ we have that

$$\lim_{N \to \infty} \frac{1}{N+1} \sum_{n=0}^{N} S_n = S. \qquad \triangle$$

4.21  THEOREM.  *If the series* $\sum_{n=0}^{\infty} a_n$ *is Abel summable to S and* $\dfrac{1}{n} \sum_{k=0}^{n} k a_k \geqslant -c$ *(for some c > 0 and n = 1, 2, ... ) then the series* $\sum_{n=0}^{\infty} a_n$ *is (C,1)-summable to S.*

*Proof.*  Let $\sigma_n = \dfrac{1}{n+1} \sum_{k=0}^{n} S_k$ *and* $f(x) = \sum_{n=0}^{\infty} a_n x^n$, $|x| < 1$. *If* $0 < x < 1$ *then*

$$f(x) = (1-x) \sum_{n=0}^{\infty} S_n x^n = (1-x)^2 \sum_{n=0}^{\infty} (n+1) \sigma_n x^n.$$

By our hypothesis we have that

$$\lim_{x \to 1-} (1-x)^2 \sum_{n=0}^{\infty} (n+1) \sigma_n x^n = S.$$

Furthermore, since $\lim_{x \to 1-} f(x) = S$, we have that

$$\lim_{x \to 1-} (1-x) \int_0^x \frac{f(t)\, dt}{(1-t)^2} = S.$$

But

$$(1-x) \int_0^x \frac{f(t)\, dt}{(1-t)^2} = (1-x) \sum_{\nu=0}^{\infty} \sigma_\nu x^{\nu+1},$$

hence

$$\lim_{x \to 1-} (1-x) \left\{ \frac{f(x)}{1-x} - \sum_{\nu=0}^{\infty} \sigma_\nu x^\nu \right\} = \lim_{x \to 1-} (1-x) \sum_{\nu=0}^{\infty} (S_\nu - \sigma_\nu) x^\nu = 0.$$

Now, $S_\nu - \sigma_\nu = \frac{1}{\nu+1} \sum_{k=1}^{\nu} k a_k \geqslant - c$ hence, by Theorem 4.20 (see (4.11) and (4.12)).

$$\sum_{\nu=0}^{n} (S_\nu - \sigma_\nu) = o(n).$$

But

$$S_\nu - \sigma_\nu = (\nu+1)\sigma_\nu - \nu \sigma_{\nu-1} - \sigma_\nu = \nu(\sigma_\nu - \sigma_{\nu-1}).$$

Writing $\sigma_{-1} = 0$ we have that $\sum_{\nu=0}^{n} \nu(\sigma_\nu - \sigma_{\nu-1}) = o(n).$

But

$$\lim_{x \to 1-} \sum_{\nu=0}^{\infty} (\sigma_\nu - \sigma_{\nu-1}) x^\nu = \lim_{x \to 1-} \frac{1-x}{x} \sum_{\nu=0}^{\infty} \sigma_\nu x^{\nu+1}$$

$$= \lim_{x \to 1-} \frac{1-x}{x} \int_0^x \frac{f(t)\, dt}{(1-t)^2} = S.$$

Hence, by Theorem 4.13 the series $\sum_{\nu=0}^{\infty} (\sigma_\nu - \sigma_{\nu-1})$ converges to $S$. Since $\sum_{\nu=0}^{n} (\sigma_\nu - \sigma_{\nu-1}) = \sigma_n$ we have that the sequence $\{\sigma_n\}_0^{\infty}$ converges to $S$, i.e. the series $\sum_{n=0}^{\infty} a_n$ is $(C,1)$-summable to $S$. $\triangle$

4.22 THEOREM (Hardy-Littlewood Theorem). *If the series* $\sum_{n=0}^{\infty} a_n$ *is Abel summable to* $S$ *and* $na_n \geqslant -c$ *(for some* $c > 0$ *and all* $n = 0, 1, \dots$ *) then the series* $\sum_{n=0}^{\infty} a_n$ *converges to* $S$.

*Proof.* Since $na_n \geqslant -c$ we have that $\frac{1}{n} \sum_{k=0}^{n} k a_k \geqslant -c$. The series $\sum_{n=0}^{\infty} a_n$ is $(C,1)$-summable to $S$ (by Theorem 4.21) and, by Theorem 4.18, the series $\sum_{n=0}^{\infty} a_n$ converges to $S$. $\triangle$

**A Tauberian Theorem for the Euler Method** $(E, \frac{1}{2})$

In Chapter 3 we defined the Euler transform, $(E,r)$, and proved that it is regular if and only if $0 < r \leqslant 1$ (see Theorem 3.15) and that $(E, s) \subset (E, \frac{1}{2})$ if $\frac{1}{2} \leqslant s \leqslant 1$ (see Theorem 3.18). Before proving the main result of this section (Theorem 4.25) we state and prove two preparatory lemmas.

4.23 LEMMA. *If* $n \geqslant 1$ *then*

$$e^{11/12} \left(\frac{n}{e}\right)^n \sqrt{n} < n! \leqslant e\left(\frac{n}{e}\right)^n \sqrt{n}.$$

*Proof.* If $0 < x < 1$ then

$$\log \frac{1+x}{1-x} = 2 \sum_{n=0}^{\infty} \frac{x^{2n+1}}{2n+1}$$

hence

$$2x < \log \frac{1+x}{1-x} < 2x + \frac{2}{3} x^2 \sum_{n=0}^{\infty} x^{2n+1} = 2x + \frac{2}{3} \frac{x^2}{1-x^2},$$

and, if $0 < x < 1$ then

$$1 < \frac{1}{2x} \log \frac{1+x}{1-x} < 1 + \frac{1}{3} \frac{x^2}{1-x^2}$$

If $x = \frac{1}{2n+1}$ $(n \geqslant 1)$ then

$$1 < (n+\tfrac{1}{2}) \log (1+\tfrac{1}{n}) < 1 + \frac{1}{12\, n(n+1)}.$$

Writing $Q_n = (n+\tfrac{1}{2}) \log [(1+(1/n)] - 1$ we have that

$$0 < Q_n < \frac{1}{12} \left(\frac{1}{n} - \frac{1}{n+1}\right).$$

For $k \geqslant 2$ we obtain

$$\log \tfrac{2}{1}(\tfrac{3}{2})^2 \cdots (\tfrac{k}{k-1})^{k-1} + \tfrac{1}{2} \log \tfrac{2}{1} \tfrac{3}{2} \cdots \tfrac{k}{k-1} = k-1 + \sum_{n=1}^{k-1} Q_n,$$

i.e.

$$\log \frac{k^k}{k!} + \tfrac{1}{2} \log k = k-1 + \sum_{n=1}^{k-1} Q_n.$$

Now

$$0 < \sum_{n=1}^{k-1} Q_n < \tfrac{1}{12} \left(1-\tfrac{1}{k}\right) < \tfrac{1}{12}$$

hence

$$e^{11/12} \left(\frac{k}{e}\right)^k \sqrt{k} < k! < e\left(\frac{k}{e}\right)^k \sqrt{k} \ (k \geqslant 2).$$

The case $n = 1$ in (4.14) is readily proved.   $\triangle$

4.24  LEMMA.  *If*

$$X_n \equiv \frac{\sqrt{n}}{2^{4n}} \sum_{k=0}^{n} \binom{4n}{k}$$

and

$$Y_n \equiv \frac{\sqrt{n}}{2^{4n}} \binom{4n}{2n}$$

*then*

$$X_n = o(1) \quad (n \to \infty) \tag{4.15}$$

*and*

$$Y_n = O(1) \quad (n \to \infty). \tag{4.16}$$

*Proof.*  To prove (4.15) we have that

$$X_n < \frac{\sqrt{n}}{2^{4n}} (n+1) \binom{4n}{n} = \frac{\sqrt{n} \ (n+1)}{2^{4n}} \frac{(4n)!}{n! \ (3n)!}$$

hence, by (4.14),

$$X_n = O\left(n \frac{2^{4n}}{3^{3n}}\right) = o(1) \quad (n \to \infty)$$

since

$$X_n < \frac{\sqrt{n} \ (n+1)}{2^{4n}} \frac{(4n)!}{n! \ (3n)!}$$

$$< \frac{\sqrt{n} \ (n+1)}{2^{4n}} \frac{e \ (4n/e)^{4n} \sqrt{(4n)}}{e^{11/12} (n/e)^n \sqrt{n} \ e^{11/12} (3n/e)^{3n} \sqrt{(3n)}}$$

$$< \frac{An^{3/2}}{2^{4n}} \frac{4^{4n}}{3^{3n} \sqrt{(3n)}} < An \frac{2^{4n}}{3^{3n}}$$

To prove (4.16) we note that

$$Y_n = \frac{\sqrt{n}}{2^{4n}} \frac{(4n)!}{(2n)!\,(2n)!}$$

$$= O\left(\frac{n}{2^{4n}} \frac{4^{4n}\, n^{4n}\, e^{2n}\, e^{2n}}{e^{4n}\, 2^{2n}\, n^{2n}\, \sqrt{n}\, 2^{2n}\, n^{2n}\, \sqrt{n}}\right) = O(1). \quad \triangle$$

**4.25 THEOREM.** *If the series* $\sum\limits_{n=0}^{\infty} a_n$ *is* $(E,\frac{1}{2})$-*summable to S and* $a_n = o(n^{-1/2})$ *then the series* $\sum\limits_{n=0}^{\infty} a_n$ *converges to S.*

*Proof.* Let $S_n = \sum\limits_{k=0}^{n} a_k$ and $\sigma_n^1 = \frac{1}{2^n} \sum\limits_{k=0}^{n} \binom{n}{k} S_k$. Since $\sum\limits_{n=0}^{\infty} a_n$ is $(E,\frac{1}{2})$-summable to $S$ we have that

$$\lim_{n\to\infty} \sigma_n^1 = S.$$

Consider

$$\sigma_{4n}^1 - S_{2n} = \frac{1}{2^{4n}} \sum\limits_{k=0}^{4n} \binom{4n}{k} (S_k - S_{2n})$$

$$= \frac{1}{2^{4n}} (\Sigma_1 + \Sigma_2 + \Sigma_3)$$

where $\quad \Sigma_1 = \sum\limits_{k=0}^{n} \binom{4n}{k} (S_k - S_{2n}), \quad \Sigma_2 = \sum\limits_{k=n+1}^{3n-1} \binom{4n}{k} (S_k - S_{2n}),$

and $\quad \Sigma_3 = \sum\limits_{k=3n}^{4n} \binom{4n}{k} (S_k - S_{2n}).$

Now, $a_n = o(n^{-1/2})$ so there exists $c > 0$ such that $|S_k| \leqslant c\sqrt{n}$ for $k \leqslant 4n$. Therefore, by Lemma 4.24,

$$\frac{1}{2^{4n}} |\Sigma_1| < \frac{2c}{2^{4n}} \sum\limits_{k=0}^{n} \binom{4n}{k} \sqrt{n} = o(1)$$

and

$$\frac{1}{2^{4n}} |\Sigma_3| < \frac{2c}{2^{4n}} \sum\limits_{k=3n}^{4n} \binom{4n}{k} \sqrt{n}$$

$$< \frac{2c}{2^{4n}} \binom{4n}{3n} (n+1) \sqrt{n} = o(1).$$

Define $\varepsilon_n = \sup_{k>n} \sqrt{k}\, |a_k|$. Since $\sqrt{n}\, a_n = o(1)$ we have that $\{\varepsilon_n\}_0^\infty$ decreases to 0. Thus

$$\frac{1}{2^{4n}}\, |\Sigma_2| = \frac{1}{2^{4n}}\, \Big| \sum_{k=n+1}^{3n-1} \binom{4n}{k}\, (S_k - S_{2n}) \Big|$$

$$\leqslant \frac{1}{2^{4n}} \sum_{k=n+1}^{3n-1} \frac{\varepsilon_n}{\sqrt{n}}\, |2n-k|\, \binom{4n}{k}$$

$$< \frac{2\varepsilon_n}{2^{4n}}\, \frac{1}{\sqrt{n}} \sum_{k=0}^{2n} (2n-k)\, \binom{4n}{k}$$

$$= \frac{2\varepsilon_n}{2^{4n}}\, \frac{1}{\sqrt{n}}\, \Big\{ 2n \sum_{k=0}^{2n} \binom{4n}{k} - 4n \sum_{k=1}^{2n} \binom{4n-1}{k-1} \Big\}$$

$$= \frac{2\varepsilon_n}{2^{4n}}\, \frac{n}{\sqrt{n}}\, \binom{4n}{2n} = 2\varepsilon_n\, Y_n$$

where $Y_n$ is defined in Lemma 4.24. By Lemma 4.24 we have that $\frac{1}{2^{4n}}\, |\Sigma_2| = o(1)$. Therefore

$$S_{2n} = S + o(1)$$

and, since $S_{2n+1} - S_{2n} = o(1)$, we have

$$S_n = S + o(1). \qquad \triangle$$

4.26  COROLLARY.  *Let* $\frac{1}{2} \leqslant r \leqslant 1$. *If* $\sum_{n=0}^\infty a_n$ *is* $(E,r)$-*summable to* $S$ *and* $a_n = o(n^{-1/2})$ *then* $\sum_{n=0}^\infty a_n$ *converges to* $S$.

*Proof.* By Theorem 3.18 $\sum_{n=0}^\infty a_n$ is $(E,\frac{1}{2})$-summable to $S$. Hence $\sum_{n=0}^\infty a_n$ converges to $S$ by Theorem 4.25. $\qquad \triangle$

# CHAPTER 5

## Fourier Series

In this chapter we present a brief account of Fourier series. We shall see, among other things, that a Fourier series is $(C,1)$-summable almost everywhere (see Theorem 5.46). A method of summation which sums the Fourier series of any function $f(\theta) \in L [-\pi,\pi)$ (i.e. $f$ is integrable, in the Lebesgue sense, on $[-\pi,\pi)$) to $f(\theta)$ almost everywhere is said to be Fourier effective. The importance of examining summability methods applied to Fourier series is seen throughout the chapter. Since summability methods come into use with respect to divergent series, it is worthwhile to note that Kolmogorov gave an example of a Fourier series which diverges almost everywhere (see N. Bary, *A Treatise on Trigonometric Series*, volume 1, pages 455-64; A. Zygmund, *Trigonometric Series*, volume I, pages 310-14; or G. H. Hardy and W. W. Rogosinski, *Fourier Series*, pages 70-2). The $(C,1)$ method and the Abel method (which is also called the Abel-Poisson method in the theory of Fourier series) are each Fourier effective whereas Cauchy convergence is not Fourier effective.

## Basic definitions and concepts

5.1 DEFINITION. *A trigonometric series is a series of the form*

$$\frac{a_0}{2} + \sum_{n=1}^{\infty} (a_n \cos nx + b_n \sin nx) \qquad (5.1)$$

*where x is a real variable and the coefficients $a_n$ and $b_n$ are constants.*
The series in (5.1) can also be written as

$$\frac{a_0}{2} + \sum_{n=1}^{\infty} (a_n \frac{e^{inx} + e^{-inx}}{2} + b_n \frac{e^{inx} - e^{-inx}}{2i}). \qquad (5.2)$$

If we let

$$c_0 = \frac{a_0}{2}, \qquad c_n = \frac{a_n - ib_n}{2},$$

and

$$c_{-n} = \frac{a_n + ib_n}{2} \quad \text{(for } n = 1, 2, \dots \text{)} \tag{5.3}$$

then expression (5.2) becomes

$$c_0 + \sum_{n=1}^{\infty} \left[ \frac{a_n - ib_n}{2} e^{inx} + \frac{a_n + ib_n}{2} e^{-inx} \right]$$

$$= c_0 + \sum_{n=1}^{\infty} c_n e^{inx} + \sum_{n=1}^{\infty} c_{-n} e^{-inx} = \sum_{n=-\infty}^{\infty} c_n e^{inx},$$

which is the complex form of the trigonometric series. If $a_n$ and $b_n$ are real then $c_{-n} = \overline{c_n}$ (recall that $\bar{z} = x - iy$ where $z = x + iy$).

5.2  DEFINITION.  *A trigonometric polynomial is a finite sum of the form*

$$s_N(x) = \frac{a_0}{2} + \sum_{n=1}^{N} (a_n \cos nx + b_n \sin nx). \tag{5.4}$$

If we integrate both sides of Eq. (5.4) from $-\pi$ to $\pi$ we obtain

$$a_0 = \frac{1}{\pi} \int_{-\pi}^{\pi} s_N(x)\,dx. \tag{5.5}$$

If we multiply both sides of Eq. (5.4) by $\cos kx$ ($\sin kx$) for $k = 1, 2, \dots, N$, and integrate from $-\pi$ to $\pi$ we obtain

$$a_k = \frac{1}{\pi} \int_{-\pi}^{\pi} s_N(x) \cos kx\,dx, \qquad k = 1, \dots, N \tag{5.6}$$

and

$$b_k = \frac{1}{\pi} \int_{-\pi}^{\pi} s_N(x) \sin kx\,dx, \qquad k = 1, \dots, N, \tag{5.7}$$

$$\text{since } \int_{-\pi}^{\pi} \cos mx \cos nx\,dx = \begin{cases} 0 & \text{if } m \neq n \\ \pi & \text{if } m = n \neq 0 \\ 2\pi & \text{if } m = n = 0, \end{cases}$$

$$\int_{-\pi}^{\pi} \sin mx \sin nx \, dx = \begin{cases} \pi & \text{if } m = n \neq 0 \\ \\ 0 & \text{otherwise,} \end{cases}$$

and

$$\int_{-\pi}^{\pi} \cos mx \sin nx \, dx = 0.$$

**5.3  DEFINITION.**  *The numbers $a_k$ and $b_k$ defined by Eqs. (5.5)–(5.7) are the Fourier coefficients of $S_N(x)$.*

In a similar fashion we have

**5.4  DEFINITION.**  *The Fourier coefficients $a_k$ and $b_k$ of a function $f \in L [-\pi, \pi)$ are given by*

$$a_k = \frac{1}{\pi} \int_{-\pi}^{\pi} f(x) \cos kx \, dx, \quad k = 0, 1, \dots, \tag{5.8}$$

and

$$b_k = \frac{1}{\pi} \int_{-\pi}^{\pi} f(x) \sin kx \, dx, \quad k = 1, 2, \dots, \tag{5.9}$$

*and the series (5.1) formed with these coefficients is the Fourier series of f.*        (*Assume f to be complex-valued unless stated otherwise.*)

We shall write

$$f(x) \simeq \frac{a_0}{2} + \sum_{n=1}^{\infty} (a_n \cos nx + b_n \sin nx) \tag{5.10}$$

to denote that the right-hand side of (5.10) is the Fourier series of $f$. We shall denote it by $S(f)$. Note that the symbol $\simeq$ implies nothing about the convergence of $S(f)$ and much less about the convergence of the series to $f(x)$. If $f$ is real-valued then $a_n$ and $b_n$ are real.

If $f$ is even, i.e. $f(-x) = f(x)$, for $x \in [-\pi, \pi)$ then $b_k = 0$ and the corresponding series is

$$f(x) \simeq \frac{a_0}{2} + \sum_{k=1}^{\infty} a_k \cos kx \text{ where } a_k = \frac{2}{\pi} \int_0^{\pi} f(x) \cos kx \, dx.$$

If $f$ is odd, i.e. $f(-x) = -f(x)$, for $x \in [-\pi, \pi)$ then $a_k = 0$ and the corresponding series is

$$f(x) \simeq \sum_{k=1}^{\infty} b_k \sin kx \quad \text{where } b_k = \frac{2}{\pi} \int_0^{\pi} f(x) \sin kx \, dx.$$

Note that $s_N(x)$ is periodic of period $2\pi$. We assume that $f$ is originally defined on $[-\pi, \pi)$ and is extended to $\underline{R}$ so as to have period $2\pi$.

For the complex form we obtain, from (5.3), (5.8), and (5.9),

$$c_k = \frac{1}{2\pi} \int_{-\pi}^{\pi} f(x) e^{-ikx} \, dx, \quad k = 0, \pm 1, \ldots$$

and

$$f(x) \simeq \sum_{k=-\infty}^{\infty} c_k e^{ikx}.$$

If $f$ is real then $c_{-k} = \overline{c_k}$.

5.5 EXAMPLE. (i) Compute the Fourier coefficients of the function where $f(x) = x, \; -\pi < x < \pi$.

We have $a_n = \frac{1}{\pi} \int_{-\pi}^{\pi} x \cos nx \, dx = 0$

and

$$b_n = \frac{1}{\pi} \int_{-\pi}^{\pi} x \sin nx \, dx = -\frac{2}{n} \cos n\pi = \frac{2}{n}(-1)^{n+1}.$$

Here the corresponding series is

$$f(x) \simeq x \simeq 2(\sin x - \frac{\sin 2x}{2} + \frac{\sin 3x}{2} + \ldots ). = S\binom{f}{0} \checkmark$$

(ii) Let $f$ be defined by

$$f(x) = \begin{cases} 0 & \text{if } x = 0 \\ \frac{1}{2}(\pi - x) & \text{if } 0 < x \leqslant \pi, \end{cases}$$

and let $f(x)$ be odd and periodic. Compute that Fourier coefficients of $f$. We have, for $n > 0$, that $b_n = \frac{2}{\pi} \int_0^\pi \frac{1}{2}(n-x) \sin nx \, dx = \frac{1}{n}$ by partial integration (note that the value of the integral is not affected by the change in the value of the function at the origin). Hence, we have

$$f(x) \simeq \sin x + \frac{\sin 2x}{2} + \frac{\sin 3x}{3} + \ldots .$$

This series is convergent for all real $x$ (see Example 1.23).

It is uniformly convergent in any closed interval not containing multiples of $2\pi$ (as is seen by Abel's inequality).

We prove now that the series is boundedly convergent, i.e.

$$| s_n(x) | = | \sum_{k=1}^{n} \frac{\sin kx}{k} | \leqslant 2 (\pi)^{1/2} \tag{5.12}$$

for all $x$. We may suppose that $0 < x < \pi$ for (5.12) is trivially satisfied when $x$ is a multiple of $\pi$ and, also, we have that $s_n(x)$ is odd and periodic (of period $2\pi$). Given $x$, let $p$ be the integer such that $p \leqslant (1/x)(\pi)^{\frac{1}{2}} < p + 1$. Thus, $| s_n(x) | \leqslant \Sigma_1 + \Sigma_2$ where

$$\Sigma_1 = \sum_{k=1}^{p} | \frac{\sin kx}{k} | \quad \text{and} \quad \Sigma_2 = | \sum_{k=p+1}^{n} \frac{\sin kx}{k} |.$$

If $p = 0$ then $\Sigma_1$ vanishes and if $p \geqslant n$ then $\Sigma_2$ vanishes. Now, $| \sin kx | \leqslant kx$ and so $\Sigma_1 \leqslant \sum_{k=1}^{p} x = px \leqslant (\pi)^{1/2}$. By Abel's inequality

$$\Sigma_2 \leqslant \frac{1}{| \sin \frac{x}{2} |} \frac{1}{p+1} = \frac{1}{\sin \frac{x}{2}} \frac{1}{p+1}$$

since $0 < x < \pi$. We now use the inequality $\sin \theta \geqslant 2\theta/\pi$ if $0 < \theta \leqslant \frac{1}{2}\pi$. (Consider $y = \frac{\sin \theta}{\theta} - \frac{2}{\pi}$. Then $\frac{dy}{d\theta}$ is negative in the interval $(0, \frac{1}{2}\pi)$ and $y(\frac{1}{2}\pi) = 0$.) Hence

$$\Sigma_2 \leqslant \frac{\pi}{x} \frac{1}{p+1} \leqslant \frac{\pi}{x} \frac{x}{(\pi)^{1/2}} = (\pi)^{1/2}$$

and (5.12) is proved.

→ *Exercise.* Expand the following functions in Fourier series.

(i) $f(x) = e^{ax}$ $(-\pi \leqslant x < \pi)$ where $a \neq 0$ is a constant

[Answer: $e^{ax} \simeq \frac{e^{a\pi} - e^{-a\pi}}{\pi}$

$[\frac{1}{2a} + \sum_{n=1}^{\infty} \frac{(-1)^n}{n^2 + a^2} (a \cos nx - n \sin nx)]]$,

(ii) $f(x) = x^2$ $(-\pi \leqslant x < \pi)$

[Answer: $x^2 \simeq \frac{\pi^2}{3} - 4(\cos x - \frac{\cos 2x}{2^2} + \frac{\cos 3x}{3^2} - \dots)]$,

(iii)  $f(x) = |x| \ (-\pi \leqslant x < \pi)$

[Answer: $|x| \simeq \dfrac{\pi}{2} - \dfrac{4}{\pi} \left( \cos x + \dfrac{\cos 3x}{3^2} + \dfrac{\cos 5x}{5^2} + \dots \right) ]$ .

## $L^2$ Space

We shall denote by $L^2 [a, b)$ the set of all functions which are measurable on $[a \ b)$ with an integrable square on $[a \ b)$.

**5.6  DEFINITION.** *Let* $f, g \in L^2 [a \ b)$. *The inner product of* $f$ *with* $g$ *is*

$$(f, g) = \int_a^b f(x) \, \overline{g(x)} \, dx,$$

*(where* $\bar{z}$ *denotes the complex conjugate of* $z$*) and the norm of* $f$ *is* $||f|| = [(f, f)]^{1/2}$.

**5.7  LEMMA.** *If* $f, g, h \in L^2 [a, b)$ *and* $\alpha$ *is a constant then*

(i)  $(f + g, h) = (f, h) + (g, h)$,

(ii)  $(f, g) = \overline{(g, f)}$,

(iii)  $(\alpha f, g) = \alpha (f, g)$,

(iv)  $||f|| \geqslant 0$ *and* $||f|| = 0$ *if and only if* $f = 0$ *almost everywhere on* $[a, b]$, *and*

(v)  $||f + g|| \leqslant ||f|| + ||g||$.

Properties (i)–(iv) follow immediately from Definition 5.6. Property (v) is Minkowski's inequality for $L^2$ (see E. Asplund and L. Bungart, *A First Course in Integration*, pages 202–3; E. C. Titchmarsh, *Theory of Functions*, page 384).  From (v) we obtain the 'triangle inequality'

$$||f - h|| \leqslant ||f - g|| + ||g - h||.$$

We write $d(f, g) = ||f - g||$.  If we agree not to distinguish between two functions, of $L^2 [a \ b)$, which coincide almost everywhere, then $L^2 [a \ b)$ (abbreviated to $L^2$) becomes a metric space with the metric (or distance function) defined by $d(f, g) = ||f - g||$.

**5.8  DEFINITION.** *Let* $F \in L^2$ *and* $F_n \in L^2$ $(n = 1, 2, \dots)$. *The sequence* $\{F_n\}_1^\infty$ *converges to* $F$ *in* $L^2$ *(written* $F_n \to F (L^2)$*) if* $||F_n - F|| = o(1)$. *The sequence* $\{F_n\}_1^\infty$ *is a Cauchy sequence in* $L^2$ *if given* $\varepsilon > 0$ *there exists a positive integer* $N = N(\varepsilon)$ *such that* $||F_m - F_n|| < \varepsilon$ *whenever* $m, n \geqslant N$.

The metric space $L^2$ is complete, that is, if $\{F_n\}_1^\infty$ is a Cauchy sequence in $L^2$, then there exists a function $F \in L^2$ such that

$||F - F_n|| = o(1)$ (see I.P. Natanson, *Theory of Functions of a Real Variable*, Vol. 1, pages 170-1).

## Orthogonal Systems

**5.9  DEFINITION.**  *A system of functions* $\phi_n \in L^2[a,b), n = 1, 2, \ldots$ *is orthogonal on* $[a,b]$ *if* $(\phi_m, \phi_n) = 0$ *for* $m \neq n$, $m = 1, 2, \ldots,$ *and* $n = 1, 2, \ldots$ *. If, in addition,* $||\phi_n|| = 1$ *for* $n = 1, 2, \ldots,$ *then the system* $\{\phi_n\}_1^\infty$ *is orthonormal.*

*Exercise.*  Prove that

$$\frac{1}{(2\pi)^{1/2}}, \frac{\cos x}{(\pi)^{1/2}}, \frac{\sin x}{(\pi)^{1/2}}, \ldots, \frac{\cos nx}{(\pi)^{1/2}}, \frac{\sin nx}{(\pi)^{1/2}}, \ldots \quad (5.13)$$

is an orthonormal system on $[-\pi, \pi]$ and, that

$$\left\{ \frac{e^{inx}}{(2\pi)^{1/2}} \right\}, \quad (n = 0, \pm 1, \pm 2, \ldots) \quad (5.14)$$

is an orthonormal system on $[-\pi, \pi]$.

**5.10  DEFINITION.**  *Let* $\{\phi_n\}_1^\infty$ *be an orthonormal system on* $[a,b]$ *and let* $f \in L^2[a,b)$. *If* $c_n = (f, \phi_n)$ *for* $n = 1, 2, \ldots$ *then* $c_n$ *is the nth Fourier coefficient of* $f$ *with respect to* $\{\phi_n\}_1^\infty$. *The Fourier series of* $f$ *with respect to* $\{\phi_n\}_1^\infty$ *is written* $f \simeq \sum_{n=1}^\infty c_n \phi_n$.

These definitions are similar to those given earlier.

## Gram-Schmidt Orthogonalization Process

**5.11  DEFINITION.**  *Let* $\{f_n\}_1^\infty$ *be a sequence of functions from* $L^2[a,b)$. *The sequence* $\{f_n\}_1^\infty$ *is linearly independent on* $[a,b]$ *provided a relation of the form* $\sum_{k=1}^n a_k f_k(x) = 0$ *almost everywhere on* $[a,b]$ *implies that* $a_1 = a_2 = \ldots = a_n = 0$.

**5.12  LEMMA.**  *Every orthonormal system,* $\{\phi_n\}_1^\infty$, *on* $[a,b]$ *is linearly independent.*

*Proof.*  Suppose that $\sum_{k=1}^n a_k \phi_k(x) = 0$ almost everywhere on $[a,b]$. Let $k$ be fixed $(1 \leqslant k \leqslant n)$. We obtain

$$\sum_{\substack{j=1 \\ j \neq k}}^n a_j \int_a^b \phi_j(x) \phi_k(x)\,dx + a_k \int_a^b |\phi_k(x)|^2\,dx = 0, \quad \text{i.e. } a_k = 0. \quad \triangle$$

Now, suppose we are given a sequence of functions $\{f_n\}_1^\infty$ where each $f_n \in L^2 [a,b)$ and the sequence is linearly independent on $[a,b]$. We can construct an orthonormal system of functions on $[a,b]$ from this given sequence in the following manner (this method is called the Gram–Schmidt process).

From properties (i)-(iii) in Lemma 5.7, we have that

$$(f + \sum_{k=1}^n a_k \phi_k, \phi_j) = (f, \phi_j) + \sum_{k=1}^n a_k (\phi_k, \phi_j).$$

Let

$$F_1(x) = f_1(x) \quad \text{and} \quad \phi_1(x) = \frac{F_1(x)}{||F_1||},$$

$$F_2(x) = f_2(x) - (f_2, \phi_1) \phi_1(x) \quad \text{and} \quad \phi_2(x) = \frac{F_2(x)}{||F_2||},$$

.
.
.

$$F_n(x) = f_n(x) - \sum_{k=1}^{n-1} (f_n, \phi_k) \phi_k(x) \quad \text{and} \quad \phi_n(x) = \frac{F_n(x)}{||F_n||},$$

.
.
.

From our construction it is readily seen that each $F_n$ and, hence, each $\phi_n$, is a linear combination of $f_1, f_2, \ldots, f_n$. Also, since the sequence $\{f_n\}_1^\infty$ is linearly independent on $[a,b]$ we have that $||F_1|| \neq 0$ and $||F_2|| \neq 0$ (for, otherwise, $F_2(x) = 0$ almost everywhere on $[a,b]$ and, thus, $f_2$ and $f_1$ are not linearly independent). Similarly, $||F_3|| \neq 0, \ldots, ||F_n|| \neq 0, \ldots$. We now show that the sequence of functions $\{\phi_n\}^\infty$ is an orthonormal system on $[a,b]$. We have

$$(\phi_n, \phi_n) = \int_a^b |\phi_n(x)|^2 \, dx = ||F_n||^{-2} \int_a^b |F_n(x)|^2 \, dx = 1.$$

To prove that $\phi_{n+1}$ is orthogonal to $\phi_1, \phi_2, \ldots, \phi_n$ we use induction. First,

$$(\phi_2, \phi_1) = \int_a^b \phi_2(x) \overline{\phi_1(x)} \, dx$$

$$= ||F_2||^{-1} \int_a^b [f_2(x) - (f_2, \phi_1) \phi_1(x)] \overline{\phi_1(x)} \, dx$$

$$= ||F_2||^{-1} [(f_2, \phi_1) - (f_2, \phi_1)] = 0.$$

Now, assume that we have $(\phi_k, \phi_j) = 0$ for $k \leqslant n$ and $j < k$. Thus, for $j \leqslant n$ we have

$$(\phi_{n+1}, \phi_j) = ||F_{n+1}||^{-1} (F_{n+1}, \phi_j)$$

and

$$(F_{n+1}, \phi_j) = (f_{n+1} - \sum_{k=1}^{n} (f_{n+1}, \phi_k) \phi_k, \phi_j)$$

$$= (f_{n+1}, \phi_j) - \sum_{k=1}^{n} (f_{n+1}, \phi_k) (\phi_k, \phi_j).$$

Since, $(\phi_k, \phi_j) = \overline{(\phi_j, \phi_k)}$ we have by our induction hypothesis that $(F_{n+1}, \phi_j) = (f_{n+1}, \phi_j) - (f_{n+1}, \phi_j) = 0$ and, hence, $(\phi_{n+1}, \phi_j) = 0$ for all $j \leqslant n$.

5.13  EXAMPLE. Let $P_n(x)$ denote the Legendre polynomial of degree $n$. The functions

$$\phi_n(x) = \left(\frac{2n+1}{2}\right)^{1/2} P_n(x), \qquad n = 0, 1, \ldots,$$

form an orthonormal system on $[-1, 1]$ (the functions $\phi_n$ are called the normed Legendre polynomials). We use Rodrigues' formula for $P_n$ (see I.P. Natanson, *Constructive Theory of Functions*, volume 2, Chapter V), namely,

$$P_n(x) = \frac{1}{n! 2^n} \frac{d^n}{dx^n} (x^2-1)^n.$$

It is easily seen that $P_0(x) = 1$, $P_1(x) = x$, and $P_2(x) = \frac{3}{2}x^2 - \frac{1}{2}$. Let $k = 0, 1, \ldots, n-1$. Therefore, we obtain

$$\int_{-1}^{1} \frac{d^n}{dx^n} (x^2-1)^n x^k \, dx = 0$$

by successive integration by parts since

$$\frac{d^{n-p}}{dx^{n-p}} (x^2-1)^n = 0 \quad \text{for } x = \pm 1 \quad \text{and} \quad 1 \leqslant p \leqslant n.$$

Therefore, $\int_{-1}^{1} P_n(x) P_m(x) dx = 0$ for $m \neq n$ and, thus, $(\phi_m, \phi_n) = 0$ for $m \neq n$.

To prove that $(\phi_n, \phi_n) = 1$ consider

$$I_n = \int_{-1}^{1} (P_n(x))^2 \, dx$$

$$= \left[ \frac{1}{n! \, 2^n} \right]^2 \int_{-1}^{1} \left[ \frac{d^n}{dx^n} (x^2-1)^n \right] \left[ \frac{d^n}{dx^n} (x^2-1)^n \right] dx.$$

By successive integration by parts we obtain

$$I_n = \frac{(-1)^n}{2^{2n}(n!)^2} \int_{-1}^{1} \frac{d^{2n}}{dx^{2n}} (x^2-1)^n \, dx = \frac{(-1)^n (2n)!}{2^{2n}(n!)^2} \int_{-1}^{1} (x^2-1)^n \, dx.$$

Since

$$\int_{-1}^{1} (x^2-1)^n \, dx = \int_{-1}^{1} (x+1)^n (x-1)^n \, dx$$

$$= [(-1)^n \, n! \, (2^{2n+1})] \, / \, [(n+1) \ldots (2n+1)]$$

we have $I_n = 2/(2n+1)$.

*Exercise.* Show that the powers $1, x, x^2, \ldots$ are linearly independent on $[-1,1]$, and, writing $f_n(x) = x^{n-1}$, $n = 1, 2, \ldots$ , compute $F_2$, $F_3$, $\phi_1$, $\phi_2$, and $\phi_3$. Verify that $\phi_1$, $\phi_2$, and $\phi_3$ are the first three normed Legendre polynomials. (Note that $a = -1$, $b = 1$ here.)

## $L_w^2$ Space

Let $w(x)$ be a non-negative function integrable $L$ on $(a, b)$, which vanishes at most at finitely many points. Let $L_w^2 [a, b)$ denote the set of all functions $f$ which are measurable on $[a, b)$ and such that $w|f|^2$ is integrable $L$ on $[a, b)$. A system of functions $\{\phi_n\}_1^\infty \in L_w^2 [a, b)$ is orthogonal on $[a, b]$ with the weight function $w$ if

$$\int_a^b w(x) \phi_n(x) \, \overline{\phi_m(x)} \, dx = 0, \quad m \neq n; \quad m, n = 1, 2, 3, \ldots .$$

If, in addition,

$$\int_a^b w(x) \, |\phi_n(x)|^2 \, dx = 1, \quad n = 1, 2, \ldots ,$$

then the system is orthonormal. The $n$th Fourier coefficient of $f \in L_w^2 [a, b)$ with respect to the orthonormal set $\{\phi_n\}_1^\infty \in L_w^2 [a, b)$ is defined to be

$$c_n = \int_a^b w(x) \, f(x) \, \overline{\phi_n(x)} \, dx, \quad n = 1, 2, \ldots ;$$

and the corresponding Fourier series is $f \simeq \sum_{n=1}^{\infty} c_n \phi_n$.

5.14  EXAMPLE.  The Chebyshev polynomials of the first kind and of degree $n$ are defined by $T_n(x) = \cos(n \arccos x)$, $n = 0, 1, \dots$ . These polynomials form an orthogonal system on $[-1,1]$ with the weight function $(1-x^2)^{-1/2}$, i.e.

$$\int_{-1}^{1} (1-x^2)^{-1/2} \cos(m \arccos x) \cos(n \arccos x)\, dx$$

$$= \int_0^{\pi} \cos m\theta \cos n\theta\, d\theta = 0 \quad \text{for } m \neq n.$$

For $m = n$ we have

$$\int_{-1}^{1} (1-x^2)^{-1/2} \cos^2(n \arccos x)\, dx = \int_0^{\pi} \cos^2 n\theta\, d\theta = \begin{cases} \frac{1}{2}\pi & \text{if } n \geqslant 1 \\ \pi & \text{if } n = 0. \end{cases}$$

Hence, the system of normed Chebyshev polynomials

$$\frac{1}{(\pi)^{1/2}}, \left(\frac{2}{\pi}\right)^{1/2} \cos(n \arccos x) \quad \text{for } n = 1, 2, \dots$$

is an orthonormal system on $[-1,1]$ with the weight function $(1-x^2)^{-1/2}$. From the formula

$$\cos n\theta = \sum_{n=0}^{[n/2]} (-1)^k \binom{n}{2k} \cos^{n-2k}\theta \sin^{2k}\theta$$

we obtain

$$T_n(x) = \sum_{k=0}^{[n/2]} (-1)^k \binom{n}{2k} x^{n-2k} (1-x^2)^k.$$

The first six Chebyshev polynomials are $T_0(x) = 1$, $T_1(x) = x$, $T_2(x) = 2x^2-1$, $T_3(x) = 4x^3-3x$, $T_4(x) = 8x^4-8x^2+1$, and $T_5(x) = 16x^5-20x^3+5x$.

*Exercise.*  (i)  Define $\hat{T}_0(x) = 1$, $\hat{T}_n(x) = (1/2^{n-1}) T_n(x)$, $n = 1, 2, \dots$ where $T_n(x)$ is given in Example 5.15.  Prove that, for $n \geqslant 1$,

(a)  $\hat{T}_n(x) = x^n + \text{terms of lower degree}$,

(b)  $\dfrac{1}{2^{n-1}} = \max_{-1 \leqslant x \leqslant 1} |\hat{T}(x)| \leqslant \max_{-1 \leqslant x \leqslant 1} |p(x)|$

where $p(x)$ is any polynomial of degree $n$ with real coefficients and leading coefficient 1.  (*Hint*: For (a) consider the expansion $\frac{1}{2}[(1+x)^n + (1-x)^n] = 1 + \binom{n}{2}x^2 + \binom{n}{4}x^4 + \dots$ .  For (b) we note that $|\hat{T}_n(x)|$ assumes its maximum value $(1/2^{n-1})$ at $n+1$ points $(x_k = \cos(k\pi/n)$ for $k = 0, 1, \ldots, n)$. If there is a polynomial $p(x) = x^n + a_n x^{n-1} + \ldots + a_n$ with $\max_{-1 \leqslant x \leqslant 1} |p(x)| < 1/2^{n-1}$ then $P(x) = \hat{T}_n(x) - p(x)$ will have alternate signs at the points $x_k$ and, so, it will have $n$ zeros; however, $P(x)$ is of degree $n-1$.)

(ii) Prove (a) $T_{n+1}(x) - 2xT_n(x) + T_{n-1}(x) = 0$

and                       (b) $(1-x^2)T_n^{(2)}(x) - xT_n^{(1)}(x) + n^2 T_n(x) = 0$.

(iii) Prove that, if $f \in L_w^2[-1,1)$, $w = (1-x^2)^{-1/2}$ then

$$f(x) \simeq \frac{a_0}{2} + \sum_{n=1}^{\infty} a_n T_n(x) \text{ where } a_n = \frac{2}{\pi} \int_{-1}^{1} (1-x^2)^{-1/2} f(x) T_n(x) \, dx.$$

## Minimal Property of Fourier Expansions

5.15  THEOREM. *Let* $f \in L^2[a,b)$. *If* $\{\phi_n\}_1^{\infty}$ *is an orthonormal system on* $[a,b]$, *then*

$$\left\| f - \sum_{k=1}^{n} (f,\phi_k)\phi_k \right\| \leqslant \left\| f - \sum_{k=1}^{n} a_k \phi_k \right\| \text{ where } a_1, \ldots, a_n \text{ are any}$$

*constants.*

*Proof.*  Consider    $(f,f)^{\frac{1}{2}} = \|f\|$

$$\left\| f - \sum_{k=1}^{n} a_k \phi_k \right\|^2 = \left( f - \sum_{k=1}^{n} a_k \phi_k, f - \sum_{k=1}^{n} a_k \phi_k \right) \quad = (f+g,h) = (f,h)+(g,h)$$

$$= \left( f, f - \sum_{k=1}^{n} a_k \phi_k \right) - \sum_{k=1}^{n} a_k \left( \phi_k, f - \sum_{j=1}^{n} a_j \phi_j \right)$$

$$= \left( f - \sum_{k=1}^{n} a_k \phi_k, f \right) - \sum_{k=1}^{n} a_k \left( f - \sum_{j=1}^{n} a_j \phi_j, \phi_k \right)$$

$$= (f,f) - \sum_{k=1}^{n} a_k (\phi_k, f) - \sum_{k=1}^{n} a_k \left[ (f,\phi_k) - \sum_{j=1}^{n} a_j (\phi_j, \phi_k) \right]$$

$$= \|f\|^2 - \sum_{k=1}^{n} \overline{a_k} (f, \phi_k) - \sum_{k=1}^{n} a_k \overline{(f,\phi_k)} + \sum_{k=1}^{n} |a_k|^2.$$

Write $c_k = (f,\phi_k)$ and add $\sum_{k=1}^{n} |c_k|^2 - \sum_{k=1}^{n} |c_k|^2$ to the right-hand side of the above expression.  Since

$$|c_k|^2 + |a_k|^2 - \overline{a_k} c_k - a_k \overline{c_k} = |c_k - a_k|^2$$

we obtain

$$||f - \sum_{k=1}^{n} a_k \phi_k ||^2 = ||f||^2 + \sum_{k=1}^{n} |c_k - a_k|^2 - \sum_{k=1}^{n} |c_k|^2. \quad (5.15)$$

Therefore,

$$||f - \sum_{k=1}^{n} c_k \phi_k ||^2 = ||f||^2 - \sum_{k=1}^{n} |c_k|^2. \quad (5.16)$$

From (5.15) and (5.16) we obtain

$$||f - \sum_{k=1}^{n} c_k \phi_k ||^2 = ||f - \sum_{k=1}^{n} a_k \phi_k ||^2 - \sum_{k=1}^{n} |c_k - a_k|^2$$

$$\leqslant ||f - \sum_{k=1}^{n} a_k \phi_k ||^2$$

and the theorem is proved. $\triangle$

Since the right-hand side of (5.15) is a minimum if and only if $a_k = c_k$ for all $k = 1, \ldots, n$ we have

**5.16 COROLLARY.** If $s_n(x) = \sum_{k=1}^{n} a_k \phi_k(x)$ then the integral $\int_a^b |f(x) - s_n(x)|^2 \, dx$ is a minimum when $s_n(x)$ is the nth partial sum of the Fourier series of $f$ with respect to $\{\phi_k\}_1^\infty$.

**5.17 COROLLARY.** (Bessel Inequality) If $f \in L^2$ then

$$\sum_{k=1}^{\infty} |c_k|^2 \leqslant ||f||^2 \quad (5.17)$$

where $c_k = (f, \phi_k)$ and $\{\phi_k\}_1^\infty$ is an orthonormal system.

*Proof.* The left-hand side of (5.16) is non-negative and so $\sum_{k=1}^{n} |c_k|^2 \leqslant ||f||^2$ for every $n = 1, 2, \ldots$ . This implies the convergence of the series $\sum_{k=1}^{\infty} |c_k|^2$ and, also (5.17). $\triangle$

**5.18 EXAMPLE.** (i) The system $1/(2\pi)^{\frac{1}{2}}$, $\cos x/(\pi)^{\frac{1}{2}}$, $\sin x/(\pi)^{\frac{1}{2}}, \ldots$ is orthonormal on $[-\pi, \pi]$ (see the Exercise following Definition 5.9). Let $f \in L^2 [-\pi, \pi)$ be real valued. Denote the Fourier coefficients by $a_k$, $b_k$ (see Eqs. 5.8) and (5.9)). We write expansion (5.10) as

$$f(x) \simeq \frac{1}{(2\pi)^{1/2}} ((2\pi)^{1/2} \frac{a_0}{2}) + \sum_{n=1}^{\infty} [ \left( \frac{\cos nx}{\pi^{\frac{1}{2}}} \right) (a_n (\pi)^{1/2})$$

$$+ \left( \frac{\sin nx}{\pi^{\frac{1}{2}}} \right) (b_n (\pi)^{1/2}) ].$$

Hence, the corresponding Bessel inequality is

$$((2\pi)^{1/2}\frac{a_0}{2})^2 + \pi \sum_{n=1}^{\infty} (a_n^2 + b_n^2) \leqslant \int_{-\pi}^{\pi} [f(x)]^2 \, dx,$$

i.e.

$$\frac{a_0^2}{2} + \sum_{n=1}^{\infty} (a_n^2 + b_n^2) \leqslant \frac{1}{\pi} \int_{-\pi}^{\pi} [f(x)]^2 \, dx.$$

(ii) The system $\dfrac{e^{inx}}{(2\pi)^{1/2}}$, $n = 0, \pm 1, \pm 2, \ldots$ is orthonormal on $[-\pi, \pi]$ (see the Exercise following Definition 5.9). Let $f \in L^2[-\pi, \pi)$. The corresponding Bessel inequality is

$$\sum_{n=-\infty}^{\infty} |c_n|^2 \leqslant \frac{1}{2\pi} \int_{-\pi}^{\pi} |f(x)|^2 \, dx$$

where

$$c_n = \frac{1}{2\pi} \int_{-\pi}^{\pi} f(t) \, e^{-int} \, dt.$$

From Bessel's inequality we deduce the following

**5.19 COROLLARY.** *The Fourier coefficients $c_n$ of $f \in L^2$ tend to 0 as $n \to \infty$.*

**5.20 THEOREM.** *The partial sums of the Fourier series of a function, $f \in L^2$, form a Cauchy sequence in $L^2$.*

*Proof.* Let $f(x) \simeq \sum_{n=1}^{\infty} c_n \phi_n(x)$ and $S_n(x) = \sum_{k=1}^{n} c_k \phi_k(x)$. For $m > n \geqslant 1$ we have

$$||S_m - S_n||^2 = ||\sum_{k=n+1}^{m} c_k \phi_k||^2 = \int_a^b |\sum_{k=n+1}^{m} c_k \phi_k(x)|^2 \, dx$$

$$= \int_a^b (\sum_{k=n+1}^{m} c_k \phi_k)(\sum_{j=n+1}^{m} c_j \phi_j) \, dx = \sum_{k=n+1}^{m} |c_k|^2.$$

By Bessel's inequality (Corollary 5.16) the series $\sum_{k=1}^{\infty} |c_k|^2$ is convergent. Therefore, given $\varepsilon > 0$ there exists a positive integer $N = N(\varepsilon)$ such that $||S_m - S_n||^2 = \sum_{k=n+1}^{m} |c_k|^2 < \varepsilon$ whenever $m \geqslant n+1$ and $n \geqslant N$. $\triangle$

We proved earlier that $f \in L^2$ and $f(x) \simeq \sum_{n=1}^{\infty} c_n \phi_n(x)$ imply $\sum_{n=1}^{\infty} |c_n|^2 < \infty$. The converse is also true.

5.21   THEOREM. (The Riesz-Fischer Theorem) *Let* $\{c_n\}_1^\infty$ *be any sequence of numbers for which* $\sum_{n=1}^\infty |c_n|^2 < \infty$ *and* $\{\phi_n\}_1^\infty$ *be any orthonormal system on* $[a,b]$. *There exists a function* $f \in L^2[a,b)$ *such that the numbers* $c_1, \ldots, c_n, \ldots$ *are its Fourier coefficients with respect to the system* $\{\phi_n\}_1^\infty$ *and such that* $S_n = \sum_{k=1}^n c_k \phi_k$ *converges to* $f$ *in* $L^2$.

*Proof.*   Consider the series $\sum_{k=1}^\infty c_k \phi_k(x)$. Since $\sum_{k=1}^\infty |c_k|^2 < \infty$, we have, given $\varepsilon > 0$, that there exists a positive integer $N = N(\varepsilon)$ such that $\sum_{k=N+1}^\infty |c_k|^2 < \varepsilon$. Hence, if $n \geqslant N$ and $p$ is any positive integer we have $\|S_{n+p} - S_n\|^2 = \sum_{k=n+1}^{n+p} |c_k|^2 < \varepsilon$, i.e. $\{S_n\}_1^\infty$ is a Cauchy sequence in $L^2$. Since $L^2$ is a complete metric space there exists a function $f \in L^2[a,b)$ such that

$$\lim_{n \to \infty} \|f - S_n\| = 0. \tag{5.18}$$

Now, for $n > k$ we have

$$(f, \phi_k) - c_k = (f, \phi_k) - (S_n, \phi_k)$$

$$= (f - S_n, \phi_k).$$

Hence, by the Cauchy–Schwarz inequality (see I. P. Natanson, *Theory of Functions of a Real Variable*, volume 1, pages 165-6) we have

$$|(f, \phi_k) - c_k| \leqslant \|f - S_n\| \, \|\phi_k\|$$

$$= \|f - S_n\| = o(1).$$

Letting $n \to \infty$, we have $c_k = (f, \phi_k)$.   $\triangle$

## Complete Orthogonal Systems

5.22   DEFINITION.   *The system of functions* $\{\phi_n\}_1^\infty$ *is complete in* $L^p[a,b)$ $(p \geqslant 1)$ *(or in* $C[a,b]$) *if there does not exist a single function* $f \in L^p[a,b)$ *(or* $f \in C[a,b]$) *which is orthogonal to all the functions of the system unless* $f = 0$ *almost everywhere on* $[a,b]$ *(for the case of* $C[a,b]$ *replace 'almost everywhere' by 'everywhere'), i.e.* $\int_a^b f(x)\phi_n(x)\,dx = 0$ *for* $n = 1, 2, \ldots$ *and* $f \in L^p[a,b)$ *imply that* $f = 0$ *almost everywhere on* $[a,b]$ *(* $f = 0$ *everywhere on* $[a,b]$ *in the case of* $C[a,b]$).

It is implied that the integrals exist, that is, if $f \in L^p [a, b)$ then $\phi_n \in L^q [a, b)$, for $n = 1, 2, \ldots$ , where $1/p + 1/q = 1$. If $f \in C [a, b]$ then $\phi_n \in L^1 [a, b)$.

Suppose now, that in the Riesz–Fischer Theorem (Theorem 5.21) the system $\{\phi_n\}_1^\infty$ is complete. If there are two functions $f, g \in L^2$ such that $c_n = (f, \phi_n)$ and $c_n = (g, \phi_n)$ for $n = 1, 2, \ldots$ then $(f - g, \phi_n) = 0$ which implies that $f(x) = g(x)$ almost everywhere on $[a, b]$. Since we identify two elements $f, g \in L^2$ as being the same if $f(x) = g(x)$ almost everywhere on $[a, b]$, the function $f$ in Theorem 5.21 is unique. Furthermore, from (5.16) and (5.18) we have

$$0 \leqslant \| f \|^2 - \sum_{k=1}^n |c_k|^2 = o(1).$$

Letting $n \to \infty$ we obtain

$$\sum_{k=1}^\infty |c_k|^2 = \| f \|^2. \qquad (5.19)$$

This relation is known as Parseval's equality. We also have

5.23  THEOREM.  *If* $\{\phi_n\}_1^\infty$ *is a complete orthonormal system,* $f \in L^2$, *and* $f \simeq \sum_{n=1}^\infty c_n \phi_n$ *then* (5.18) *and* (5.19) *hold.*

We shall prove later in this chapter that the systems (5.13) and (5.14) are complete in $L [-\pi, \pi)$.

## Mercer's Theorem

We now suppose that $f \in L$ and that the functions $\{\phi_n\}_1^\infty$ are bounded.

5.24  THEOREM. (Mercer's Theorem) *If* $\{\phi_n\}_1^\infty$ *is an orthonormal system and* $|\phi_n(x)| \leqslant M$ *(for some* $M > 0$*) for all* $n = 1, 2, \ldots$ *and all* $a \leqslant x \leqslant b$ *then the Fourier coefficients of any function* $f \in L [a, b)$ *tend to zero.*

*Proof.*  Let $\varepsilon > 0$ be given. Since $f \in L [a, b)$ there exists a bounded function $B$ such that $\int_a^b |f(x) - B(x)| \, dx < \varepsilon$. Since $B \in L^2 [a, b)$ we have, by Corollary 5.16, that there exists a positive integer $N = N(\varepsilon)$ such that if $n \geqslant N$ then $|\int_a^b B(x) \overline{\phi_n(x)} \, dx| < \varepsilon$. Therefore,

$$|c_n| = |\int_a^b f(x) \overline{\phi_n(x)} \, dx | = |\int_a^b [(f(x) - B(x))\overline{\phi_n(x)} + B(x)\overline{\phi_n(x)}] \, dx |$$

$$\leqslant |\int_a^b (f(x) - B(x))\overline{\phi_n(x)} \, dx | + |\int_a^b B(x)\overline{\phi_n(x)} \, dx |$$

$$\leqslant M\varepsilon + \varepsilon \quad \text{for } n \geqslant N. \quad \triangle$$

**5.25 COROLLARY.** (Riemann–Lebesgue Theorem). *If $f \in L\ [-\pi,\pi)$ then the Fourier coefficients $a_n$, $b_n$ with respect to system (5.13) and the Fourier coefficients $c_n$ with respect to system (5.14) each tend to zero.*

## Dirichlet's Integral

Let $f \in L\ [-\pi,\pi)$ and suppose that $f$ has period $2\pi$. If $a_k$, $b_k$ are the Fourier coefficients of $f$ then, for $-\pi \ll x \ll \pi$, we have

$$a_k = \frac{1}{\pi} \int_{-\pi}^{\pi} f(t) \cos kt \, dt = \frac{1}{\pi} \int_{x-\pi}^{x+\pi} f(t) \cos kt \, dt, \quad k = 0, 1, \dots ;$$

for

$$\int_{x-\pi}^{-\pi} f(t) \cos kt \, dt = \int_{x+\pi}^{\pi} f(t) \cos kt \, dt \text{ by periodicity.}$$

Similarly

$$b_k = \frac{1}{\pi} \int_{-\pi}^{\pi} f(t) \sin kt \, dt = \frac{1}{\pi} \int_{x-\pi}^{x+\pi} f(t) \sin kt \, dt, \quad k = 1, 2, \dots ;$$

and

$$S_n(x) = \frac{a_0}{2} + \sum_{k=1}^{n} (a_k \cos kx + b_k \sin kx)$$

$$= \frac{1}{\pi} \int_{x-\pi}^{x+\pi} \frac{f(t)}{2} \, dt$$

$$+ \sum_{k=1}^{n} \left[ \frac{1}{\pi} \int_{x-\pi}^{x+\pi} f(t) \, (\cos kt \cos kx + \sin kt \sin kx) \, dt \right]. \qquad (5.20i)$$

So,

$$S_n(x) = \frac{1}{\pi} \int_{x-\pi}^{x+\pi} f(t) \left[ \frac{1}{2} + \sum_{k=1}^{n} \cos k\,(t-x) \right] dt.$$

Let $D_n(x) = \frac{1}{2} + \sum_{k=1}^{n} \cos kx$ (see Example 1.23). Thus

$$S_n(x) = \frac{1}{\pi} \int_{x-\pi}^{x+\pi} f(t) \, \frac{\sin (n+\frac{1}{2})\,(t-x)}{2 \sin ((t-x)/2)} \, dt.$$

Now, let $t-x = u$. We obtain

$$S_n(x) = \frac{1}{\pi} \int_{-\pi}^{\pi} f(u+x) \, \frac{\sin (n+\frac{1}{2})\,u}{2 \sin (u/2)} \, du$$

$$= \frac{1}{\pi} \left[ \int_{-\pi}^{0} f(u+x) \, \frac{\sin (n+\frac{1}{2})\,u}{2 \sin (u/2)} \, du + \int_{0}^{\pi} f(u+x) \, \frac{\sin (n+\frac{1}{2})\,u}{2 \sin (u/2)} \, du \right].$$

Hence,

$$S_n(x) = \frac{1}{\pi} \int_0^\pi \left( f(x+u) + f(x-u) \right) \frac{\sin\ (n+\frac{1}{2})\ u}{2\ \sin\ (u/2)}\ du \qquad (5.20\text{ii})$$

Formula (5.20ii) is known as *Dirichlet's integral.* If we take $f(x) = 1$ then we obtain $a_0/2 = 1$, $a_n = 0 = b_n$, $S_n(x) = 1$, and, consequently, we have (from (5.20ii))

$$1\ =\ \frac{1}{\pi} \int_0^\pi 2\ \frac{\sin\ (n+\frac{1}{2})\ u}{2\ \sin\ (u/2)}\ du.$$

Multiplying this formula by $s$ and subtracting it from (5.20ii) we obtain

$$S_n(x) - s = \frac{1}{\pi} \int_0^\pi \left( f(x+u) + f(x-u) - 2s \right) \frac{\sin\ (n+\frac{1}{2})\ u}{2\ \sin\ (u/2)}\ du. \qquad (5.21)$$

We now require the following

**5.26  LEMMA.** *Let $f, g \in L\ [-\pi, \pi)$ and let both functions have period $2\pi$. If $g$ is bounded and $x$ is real then*

$$c = c(x, \lambda)\ =\ \int_{-\pi}^\pi f(x+t)\ g\ (t)\ \cos\ \lambda t\ dt$$

*and*

$$s = s(x, \lambda)\ =\ \int_{-\pi}^\pi f(x+t)\ g\ (t)\ \sin\ \lambda t\ dt \qquad (5.22)$$

*both tend to zero uniformly in $x$ as $\lambda \to \infty$.*

*Proof.* Write $\psi_x(t) = f(x+t)\ g\ (t)$ and $M = \sup_t\ |g(t)|$. Now,

$$c\ =\ \int_{-\pi}^\pi \psi_x\ (T)\ \cos\ \lambda t\ dT$$

$$=\ -\int_{-\pi-\pi/\lambda}^{-\pi} \psi_x\ (t+\frac{\pi}{\lambda})\ \cos\ \lambda t\ dt - \int_{-\pi}^\pi \psi_x\ (t+\frac{\pi}{\lambda})\ \cos\ \lambda t\ dt$$

$$-\int_\pi^{\pi-\pi/\lambda} \psi_x\ (t+\frac{\pi}{\lambda})\ \cos\ \lambda t\ dt.$$

Since $\psi_x$ has period $2\pi$ we have that

$$c\ =\ -\int_{-\pi}^\pi \psi_x\ (t+\frac{\pi}{\lambda})\ \cos\ \lambda t\ dt. \qquad (5.23)$$

From (5.22) and (5.23) we obtain

$$|c| = |\frac{1}{2} \int_{-\pi}^{\pi} [\psi_x(t) - \psi_x(t+\frac{\pi}{\lambda})] \cos \lambda t \, dt |$$

$$\leqslant \frac{1}{2} \int_{-\pi}^{\pi} | \psi_x(t+\frac{\pi}{\lambda}) - \psi_x(t) | \, dt$$

$$= \frac{1}{2} \int_{-\pi}^{\pi} | (f(x+t+\frac{\pi}{\lambda}) - f(x+t)) g(t+\frac{\pi}{\lambda}) + f(x+t)(g(t+\frac{\pi}{\lambda}) - g(t)) | \, dt$$

$$\leqslant I_1 + I_2$$

where

$$I_1 = \frac{1}{2} \int_{-\pi}^{\pi} | f(x+t+\frac{\pi}{\lambda}) - f(x+t) || g(t+\frac{\pi}{\lambda})| \, dt$$

and

$$I_2 = \frac{1}{2} \int_{-\pi}^{\pi} | g(t+\frac{\pi}{\lambda}) - g(t) || f(x+t) | \, dt.$$

So,

$$I_1 \leqslant \frac{M}{2} \int_{-\pi}^{\pi} | f(x+t+\frac{\pi}{\lambda}) - f(x+t) | \, dt = \frac{M}{2} \int_{-\pi}^{\pi} | f(t+\frac{\pi}{\lambda}) - f(t) | \, dt$$

$$\leqslant \frac{M}{2} \omega_1(\frac{\pi}{\lambda}, f) \text{ where } \omega_1(\delta, f) = \sup_{0 \leqslant |h| \leqslant \delta} \{ \int_{-\pi}^{\pi} | f(t+h) - f(t) | \, dt \}$$

($\omega_1(\delta, f)$ is called the integral modulus of continuity of $f$). Since $f \in L [-\pi, \pi)$ and is periodic we have $\omega_1(\delta, f) \to 0$ as $\delta \to 0$ (see A. Zygmund, *Trigonometric Series*, Volume 1, Chapter 2; N. Bary, *A Treatise on Trigonometric Series*, Volume 1, pages 37–41).

We now estimate $I_2$. Given $\varepsilon > 0$ we can write $f = f_1 + f_2$ where $f_1$ is bounded (say $|f_1(t)| \leqslant K$) and $\int_{-\pi}^{\pi} |f_2(t)| \, dt < \varepsilon$. Therefore

$$I_2 \leqslant \frac{K}{2} \int_{-\pi}^{\pi} |g(t+\frac{\pi}{\lambda}) - g(t)| \, dt + \frac{1}{2} \int_{-\pi}^{\pi} |f_2(x+t)| \, |g(t+\frac{\pi}{\lambda}) - g(t)| \, dt$$

$$< \frac{K}{2} \omega_1(\frac{\pi}{\lambda}, g) + M\varepsilon.$$

Hence, there exists $\lambda_0(\varepsilon, M, K)$ such that

$$|c| < \frac{M}{2} \omega_1(\frac{\pi}{\lambda}, f) + \frac{K}{2} \omega_1(\frac{\pi}{\lambda}, g) + M\varepsilon < 2M\varepsilon \quad \text{for } \lambda \geqslant \lambda_0(\varepsilon, M, K).$$
$$(5.24)$$

Since $\varepsilon$ is arbitrary and the right-hand side of (5.24) is independent of $x$ we have that $c \to 0$ uniformly as $\lambda \to \infty$.

The proof for $s$ is similar.    △

5.27  COROLLARY.  *The statement of Lemma 5.26 holds for the integrals* $\int_a^b f(x+t) g(t) \cos \lambda t \, dt$ *and* $\int_a^b f(x+t) g(t) \sin \lambda t \, dt$ *where a and b are any two points in* $[-\pi, \pi]$.

*Proof.*  We can reduce this case to the preceding one by taking

$$g_1(t) = \begin{cases} g(t) & \text{in } [a, b) \\ \\ 0 \text{ elsewhere} & \text{in } [-\pi, \pi). \quad △ \end{cases}$$

We also obtain from Lemma 5.26

5.28  COROLLARY.  (Riemann–Lebesgue Theorem)  *If* $g \in L \ [-\pi, \pi)$ *then* $\lim_{\lambda \to \infty} \int_{-\pi}^{\pi} g(t) \cos \lambda t \, dt = 0$  *and*  $\lim_{\lambda \to \infty} \int_{-\pi}^{\pi} g(t) \sin \lambda t \, dt = 0$.

This corollary implies that the Fourier coefficients of $g$ are $o(1)$, a result obtained earlier with the help of Mercer's Theorem.

### Convergence of Fourier Series

We assume that $f \in L \ [-\pi, \pi)$ and that $f$ has period $2\pi$.  Let $0 < \delta < \pi$.          Fix $x$ and a complex number $s$.          Define $\phi(u) = f(x+u) + f(x-u) - 2s$.  Thus $\phi \in L \ [-\pi, \pi]$.  From (5.21) we have

$$S_n(x) - s = \frac{1}{\pi} \left[ \int_0^{\delta} \phi(u) \, \frac{\sin \ (n+\frac{1}{2}) \, u}{2 \sin \ (u/2)} \, du \right.$$

$$\left. + \int_{\delta}^{\pi} \phi(u) \, \frac{\sin \ (n+\frac{1}{2}) \, u}{2 \sin \ (u/2)} \, du \right]. \tag{5.25}$$

Note that $\phi(u)/\sin(u/2)$ is integrable on $[\delta, \pi)$ so, by Corollary 5.28, the second integral in Eq. (5.25) is $o(1)$ as $n \to \infty$.

For the first integral in Eq. (5.25) we write

$$\frac{\sin \ (n+\frac{1}{2}) \, u}{2 \sin \ (u/2)} = \frac{\sin \ nu}{u} + \left( \frac{1}{2 \tan \ (u/2)} - \frac{1}{u} \right) \sin \ nu + \frac{1}{2} \cos \ nu.$$

Now, $\int_0^{\delta} \phi(u) \cos nu \, du = o(1)$.  Since $u - 2 \tan(u/2) = O(u^3)$ as $u \to 0$ the function

$$g(u) = \frac{1}{2 \tan \ (u/2)} - \frac{1}{u}$$

is continuous on $[0,\delta]$. Hence,

$$\int_0^\delta \phi(u) \left( \frac{1}{2 \tan (u/2)} - \frac{1}{u} \right) \sin nu \; du = o(1),$$

and the right-hand side of (5.25) reduces to

$$S_n(x) - s = \frac{1}{\pi} \int_0^\delta \phi(u) \frac{\sin nu}{u} du + o(1). \qquad (5.26)$$

Take $s = 0$. Thus (5.26) can be written as

$$S_n(x) = \frac{1}{\pi} \int_{-\delta}^\delta f(x+u) \frac{\sin nu}{u} du + o(1).$$

Since the integral utilizes the values of $f$ in $(x-\delta, x+\delta)$ we have proved

**5.29 THEOREM.** (Riemann's Principle of Localization)
*The convergence or divergence of the Fourier series at a point $x$ depends only upon the behaviour of the function $f$ in the immediate neighborhood of the point $x$.*

### Convergence Tests
We use formula (5.26) for $S_n(x) - s$ to obtain the following necessary and sufficient condition for convergence.

**5.30 THEOREM.** *The Fourier series of a function $f \in L [-\pi, \pi)$ converges, at a point $x$, to the value $s$ if and only if* $\lim_{n\to\infty} \int_0^\delta \frac{\phi(u)}{u} \sin nu \; du = 0$ *for some $\delta \in (0, \pi)$.*

The following sufficient condition for convergence follows immediately from Theorem 5.30 and Corollary 5.28.

**5.31 THEOREM.** (Dini's Test) *If $\phi(u)/u \in L [0,\delta)$ for some $\delta > 0$ then $\lim_{n\to\infty} S_n(x) = s$.*

**5.32 COROLLARY.** *If $f$ is differentiable at $x$ then the series converges at $x$ to $f(x)$.*

*Proof.* Let $s = f(x)$ and note that

$$\lim_{u\to 0} \left( \frac{f(x+u) - f(x)}{u} \right) = \lim_{u\to 0} \left( \frac{f(x-u) - f(x)}{-u} \right) = f^{(1)}(x).$$

In a neighborhood of $x$ these two expressions in parentheses, and consequently $\phi(u)/u$, are bounded functions of $u$. Therefore, the corollary holds by applying Dini's test (Theorem 5.31). $\triangle$

In the following corollary we consider functions $f$ satisfying the condition

$$|f(x + h) - f(x)| \leqslant c |h|^{\alpha}, \quad \alpha > 0, \quad c > 0,$$

at a point $x$. The function $f$ is said to satisfy Lipschitz condition of order $\alpha$ at the point $x$. If this condition is satisfied for all $x$ in the interval $[-\pi, \pi)$, we write $f \in \mathrm{Lip}_c \alpha$.

**5.33 COROLLARY.** *If $f$ satisfies a Lipschitz condition of order $\alpha$, $0 < \alpha$, at the point $x$ then the series converges at $x$ to $f(x)$.*

*Proof.* Suppose that $|f(x+h) - f(x)| \leqslant c |h|^{\alpha}$ for some $\alpha > 0$. Therefore,

$$\left| \frac{\phi(u)}{u} \right| = \left| \frac{f(x+u) + f(x-u) - 2f(x)}{u} \right| \leqslant 2cu^{\alpha-1}$$

and the corollary is proved by applying Dini's test (Theorem 5.31) with $s = f(x)$. $\triangle$

**5.34 THEOREM.** (Jordan's Test) *If $f$ is a function of $\underline{bounded}$ $\underline{variation}$ on the interval $[a, b]$ then its Fourier series converges at $\underline{every\ point}$ $x \in (a, b)$ to $\frac{1}{2}(f(x+0) + f(x-0))$ where $f(x+0)$ and $f(x-0)$ are the right and left-hand limits at $x$ respectively.*

For the proof of Theorem 5.34 we require the following two lemmas (after which we will supply proof of the theorem).

**5.35 LEMMA.** (Second Mean Value Theorem) *If $f \in L [a, b)$, real-valued, and $\phi$ is a non-negative, non-decreasing function of $[a, b]$ then there is a $\xi \in [a, b]$ such that $\int_a^b \phi(x) f(x)\, dx = \phi(b-0) \int_\xi^b f(x)\, dx$. If $\phi$ is non-negative and non-increasing then there exists a $\eta \in [a, b]$ such that $\int_a^b \phi(x) f(x)\, dx = \phi(a+0) \int_a^\eta f(x)\, dx$.*

For a proof of this lemma see E.C. Titchmarsh, *Theory of Functions*, pages 379–81 or see E. Asplund and L. Bungart, *A First Course in Integration*, pages 432–33.

**5.36 LEMMA.** *If $0 \leqslant a \leqslant b$ then $\left| \int_a^b \frac{\sin t}{t}\, dt \right| < \pi$.*

*Proof.* (i) If $a \geqslant 4/\pi$ then we obtain, by Lemma 5.35 with $\phi(x) = 1/x$, that there exists $\eta \in [a, b]$ such that

$$\left| \int_a^b \frac{\sin x}{x} \, dx \right| = \left| \frac{1}{a} \int_a^\eta \sin x \, dx \right| \leqslant \frac{2}{a} \leqslant \frac{\pi}{2}.$$

(ii) If $a < 4/\pi$ and $b \geqslant 4/\pi$ then

$$\int_a^b \frac{\sin x}{x} \, dx = \int_a^{4/\pi} \frac{\sin x}{x} \, dx + \int_{4/\pi}^b \frac{\sin x}{x} \, dx = I_1 + I_2.$$

For $I_1$ we have $\sin x \leqslant x$ and, thus, $I_1 \leqslant 4/\pi < \frac{1}{2}\pi$. For $I_2$ we have, by part (i), that $|I_2| \leqslant \frac{1}{2}\pi$. Consequently, $|I_1 + I_2| < \pi$.

(iii) If $b < 4/\pi$ then

$$\int_a^b \frac{\sin x}{x} \, dx \leqslant b < \frac{4}{\pi} < \frac{1}{2}\pi. \qquad \triangle$$

We are now ready to prove Theorem 5.34.

*Proof* (of Theorem 5.34). Let $x \in (a,b)$. It is enough to prove the theorem when $f$ is real-valued for if $f = f_1 + if_2$ then $f_1$, $f_2 \in BV[a,b]$ and are real-valued. Since $f \in BV[a,b]$ ($f$ is of bounded variation over $[a,b]$) we have that the limits $f(x+0)$ and $f(x-0)$ exist. Let $s = \frac{1}{2}(f(x+0) + f(x-0))$. Therefore, $\phi(u) = f(x+u) + f(x-u) - f(x-0) - f(x-0)$ is of bounded variation on $[0,\lambda]$ where $0 < \lambda \leqslant \min(\pi, x-a, b-x)$ and $\phi(u) \to 0$ as $u \to 0^+$ (from the right). Hence, we can write $\phi(u) = \phi_1(u) - \phi_2(u)$ where $\phi_j \geqslant 0$ and is increasing ($j = 1, 2$). Since each $\phi_j$ tends to the same limit as $u \to 0^+$ we may suppose, by subtracting a suitable constant if necessary, that $\lim_{u \to 0^+} \phi_j(u) = 0$ ($j = 1, 2$). By Theorem 5.30 it suffices to show that $\lim_{n \to \infty} I_j = 0$ ($j = 1, 2$) where $I = \int_0^\lambda \frac{\phi_j(u)}{u} \sin nu \, du$.

Consider the integral $I_1$. Given $\varepsilon > 0$ choose $\eta$, $0 < \eta < \lambda$, such that $\phi(\eta) < \varepsilon/\pi$. Thus, by the second mean value theorem (Lemma 5.35), there exists $\xi \in [0,\eta]$ such that

$$\left| \int_0^\eta \frac{\phi_1(u)}{u} \sin nu \, du \right| = \left| \phi_1(\eta) \int_\xi^\eta \frac{\sin nu}{u} \, du \right|$$

$$= \left| \phi_1(\eta) \int_{n\xi}^{n\eta} \frac{\sin u}{u} \, du \right| < \pi \frac{\varepsilon}{\pi} = \varepsilon.$$

Fix $\eta$. By the Riemann-Lebesgue theorem (Corollary 5.28) we have that there exists $n_1$ such that $\left| \int_\eta^\lambda [\phi_1(u)/u] \sin nu \, du \right| < \varepsilon$ for $n \geqslant n_1$.

Thus $|I_1| < 2\varepsilon$ for $n \geqslant n_1$.    Similarly, there exists $n_2$ such that $|I_2| < 2\varepsilon$ for $n \geqslant n_2$ and, hence, the theorem is proved.    $\triangle$

*Exercise.*   (i)   Suppose that $f \in BV[a,b]$. If $[a',b'] \subseteq (a,b)$ prove that the partial sums $S_n(x)$ remain bounded on $[a',b']$.

     (ii)   Let $f \in BV[a,b]$.    If $f$ is continuous on $[a,b]$ and $[a',b'] \subseteq (a,b)$ prove that the Fourier series of $f$ converges uniformly on $[a',b']$.

## Cesàro Summability of Fourier Series

     Let $f \in L[-\pi,\pi)$ and suppose that $f$ has period $2\pi$.   In the previous sections we discussed conditions which ensure the convergence of the Fourier series of $f$ at a given point.   We will now consider the $(C,1)$ summability of the Fourier series.   We write (see Example 1.23)

$$D_n(u) = \tfrac{1}{2} + \sum_{k=1}^{n} \cos ku \text{ and } K_n(u) = \frac{1}{n+1}(D_0(u) + \ldots + D_n(u)).$$

Since $2 \sin (k+\tfrac{1}{2})u \sin (u/2) = \cos ku - \cos (k+1)u$, we obtain

$$K_n(u) = \begin{cases} \dfrac{1-\cos(n+1)u}{4(n+1)\sin^2(u/2)} & \text{if } u \neq 0 \;(\mathrm{mod}\; 2\pi) \\[3mm] (n+1)/2 & \text{if } u = 0 \;(\mathrm{mod}\; 2\pi). \end{cases}$$

     The functions $D_n$ are called Dirichlet kernels and $K_n$ are called Fejér kernels.   One can easily prove

(i)                      $K_n(-u) = K_n(u),$

(ii)                   $|D_n(u)| \leqslant n+\tfrac{1}{2},$

(iii)                $0 \leqslant K_n(u) \leqslant \tfrac{1}{2}(n+1),$    and

(iv)       $\dfrac{1}{\pi}\int_{-\pi}^{\pi} D_n(u)\,du = 1,$    $\dfrac{1}{\pi}\int_{-\pi}^{\pi} K_n(u)\,du = 1.$      (5.27)

     Since $\sin \theta \geqslant 2\theta/\pi$ for $0 \leqslant \theta \leqslant \tfrac{1}{2}\pi$ we obtain

(v)          $K_n(u) \leqslant \pi^2/2\,(n+1)\,u^2$    for $0 < |u| \leqslant \pi.$      (5.27)

It follows from (5.27,v) that

(vi)         $\lim_{n\to\infty}\int_{\delta}^{\pi} K_n(u)\,du = 0$    for $0 < \delta < \pi.$      (5.27)

Now, let

$$\sigma_n(x) = \sigma_n(x,f) = \frac{1}{n+1}(S_0(x) + \ldots + S_n(x)). \qquad (5.28)$$

Then, from (5.20,i) and (5.28), we have

$$\sigma_n(x) = \frac{a_0}{2} + \sum_{k=1}^{n}(a_k \cos kx + b_k \sin kx)(1 - \frac{1}{n+1}); \qquad (5.29)$$

and, from (5.20,ii), we obtain

$$\sigma_n(x) = \frac{1}{\pi}\int_0^{\pi}(f(x+u) + f(x-u))\left(\frac{D_0(u) + \ldots + D_n(u)}{n+1}\right)du$$

$$= \frac{1}{\pi}\int_0^{\pi}(f(x+u) + f(x-u))K_n(u)\,du. \qquad (5.30)$$

Since $K_n(u)$ is even we have, from (5.27,iv),

$$1 = \frac{2}{\pi}\int_0^{\pi}K_n(u)\,du. \qquad (5.31)$$

Multiplying (5.31) by $s$ and subtracting from (5.30) gives us

$$\sigma_n(x) - s = \frac{1}{\pi}\int_0^{\pi}(f(x+u) + f(x-u) - 2s)K_n(u)\,du. \qquad (5.32)$$

We are now ready to prove

see pg 130 of Bartle

5.37  THEOREM. (Fejér) (i) *If, for some* $x$, $f(x+0)$ *and* $f(x-0)$ *exist then the Fourier series of* $f$ *in* $(C,1)$ *summable at the point* $x$ *to the sum*

$$\frac{f(x+0) + f(x-0)}{2}.$$

(ii)  *If* $f \in C[a,b]$ *and* $[a',b'] \subseteq (a,b)$ *then* $\lim_{n\to\infty}\sigma_n(x) = f(x)$ *uniformly on* $[a',b']$.

(iii)  *If* $f \in C[-\pi,\pi]$ *then* $\lim_{n\to\infty}\sigma_n(x) = f(x)$ *uniformly on* $[-\pi,\pi]$.

(*Remark.*  The hypothesis of (iii) implies, by periodicity, that $f(-\pi) = f(\pi)$.  A similar remark applies to (ii) if $[-\pi,\pi] \subseteq [a, b]$.)

*Proof.* (i)  Let $s = \frac{1}{2}(f(x+0) + f(x-0))$ in (5.32).     Therefore $\phi(u) = f(x+u) + f(x-u) - f(x+0) - f(x-0)$ and, hence $\phi(u) \to 0$ as $u \to 0$. Hence, given $\varepsilon > 0$, there exists $\delta > 0$ such that $|\phi(u)| \leqslant \varepsilon$ for $|u| \leqslant \delta$.

We now write Eq. (5.32) as

$$\sigma_n(x) - s = I_1 + I_2$$

where

$$I_1 = \frac{1}{\pi} \int_0^{\delta} \phi(u) K_n(u) \, du$$

and

$$I_2 = \frac{1}{\pi} \int_{\delta}^{\pi} \phi(u) K_n(u) \, du.$$

So

$$|I_1| < \frac{\varepsilon}{\pi} \int_0^{\delta} K_n(u) \, du < \frac{\varepsilon}{\pi} \frac{\pi}{2} = \frac{\varepsilon}{2}$$

and

$$|I_2| < \frac{1}{\pi} \frac{\pi^2}{2(n+1)\delta^2} \int_{\delta}^{\pi} |\phi(u)| \, du < \frac{c(\delta)}{n+1}$$

where $c(\delta)$ is a constant depending upon $\delta$ (since $\phi(u) \in L \ [\delta, \pi)$). Hence $\lim_{n \to \infty} |\sigma_n - s| < \frac{1}{2}\varepsilon$ and the first part of the theorem is proved.

(ii)  Let $\lambda = \min(\pi, b-b', a'-a)$. Since $f$ is continuous on the closed interval $[a,b]$, given $\varepsilon > 0$, we can choose $\delta = \delta(\varepsilon) \leqslant \lambda$ such that for $u \leqslant \delta$ and $a' \leqslant x \leqslant b'$ we have $|f(x+u) - f(x)| < \varepsilon$ and $|f(x-u) - f(x)| < \varepsilon$. Take $s = f(x)$. If $u \leqslant \delta$ and $a' \leqslant x \leqslant b'$ then $|\phi(u)| < 2\varepsilon$. As before we write $\sigma_n(x) - s = I_1 + I_2$ where $|I_1| < \varepsilon$. Similarly, there exists $n_0(\varepsilon) > 0$ such that $|I_2| < (\pi/[2(n+1)\delta^2] \ c(\delta) < \varepsilon$ for $n \geqslant n_0(\varepsilon)$ and $a' \leqslant x \leqslant b'$. Hence, $\sigma_n(x)$ converges uniformly to $f(x)$ on $[a', b']$.

(iii)  By periodicity, $f(x)$ is continuous on any interval $[A,B]$ and the result follows from (ii).  $\triangle$

5.38  COROLLARY.  *Suppose, for some* $x$, $f(x+0)$ *and* $f(x-0)$ *exist. If* $\{S_n(x)\}$ *converges then it must converge to* $\frac{1}{2}(f(x+0) + f(x-0))$.

*Proof.*  The series is $(C,1)$ summable to $\frac{1}{2}(f(x+0) + f(x-0))$. Since the $(C,1)$ method is regular, $S_n(x)$ must converge to the $(C,1)$ sum, i.e. $S_n(x)$ converges to $\frac{1}{2}(f(x+0) + f(x-0))$.  $\triangle$

5.39  COROLLARY. (Weierstrass Approximation Theorem 1) *Let* $f \in C \ [-\pi,\pi]$ *and have period* $2\pi$.     *Given* $\varepsilon > 0$ *there exists a trigonometric polynomial* $T(x)$ *such that* $|f(x) - T(x)| < \varepsilon$ *for all* $x$.
       The proof follows from Theorem 5.37 (iii), for each $\sigma_n(x)$ is a trigonometric polynomial (see (5.29)).

**5.40  COROLLARY.** (Weierstrass Approximation Theorem 2) *Let* $f \in C\,[a,b]$. *Given* $\varepsilon > 0$ *there exists a polynomial* $p\,(x)$ *such that* $|f(x) - p\,(x)| < \varepsilon$ *for all* $a \leqslant x \leqslant b$.

*Proof.* We may suppose that $[a,b] = [0,\pi]$ (otherwise, consider the function $f[\pi(x-a)/(b-a)]$). We extend $f(x)$ to $(-\pi,0)$ by $f(x) = f(-x)$ for $-\pi < x < 0$, and, then by periodicity so that $f$ is continuous and periodic. By Theorem 5.37 (iii) we can find $n_0(\varepsilon) > 0$ such that for $n \geqslant n_0(\varepsilon)$ we have $|f(x) - \sigma_n(x)| < \frac{1}{2}\varepsilon$ for all $0 \leqslant x \leqslant \pi$. Each term in $\sigma_n(x)$ is a linear combination of $\cos kx$ (see (5.29); $b_k = 0$ since $f$ is even). We have that $\cos kx = \sum\limits_{j=0}^{\infty} (-1)^j\,[(kx)^{2j}/(2j)!]$ the convergence being uniform on $[a,b]$; therefore, we can replace each $\cos kx$ in $\sigma_n(x)$ by a polynomial $p_k(x)$ and, thus, obtain a polynomial $p(x)$ such that $|\sigma_n(x) - p(x)| < \frac{1}{2}\varepsilon$ for all $0 \leqslant x \leqslant \pi$. Hence, $|f(x) - p(x)| < \varepsilon$ for all $0 \leqslant x \leqslant \pi$.    $\triangle$

**5.41  EXAMPLE.** If $f(x)$ is real-valued and $m \leqslant f(x) \leqslant M$ for all $x$ then $m \leqslant \sigma_n(x) \leqslant M$ for all $n = 0, 1, \ldots$ . From (5.30) we see that

$$\sigma_n(x) \leqslant \frac{2M}{\pi} \int_0^{\pi} K_n(u)\,du = M$$

and that

$$\sigma_n(x) \geqslant \frac{2M}{\pi} \int_0^{\pi} K_n(u)\,du = m.$$

The next two examples give an upper bound to $|f(x) - \sigma_{n-1}(x)|$.

**5.42  EXAMPLE.** If a $2\pi$-period function $f \in \text{Lip}_c\alpha$, $0 < \alpha < 1$ then for all real $x$

$$||f - \sigma_{n-1}||_c \equiv \sup\nolimits_{-\pi \leqslant x \leqslant \pi} |f(x) - \sigma_{n-1}(x)| \leqslant \frac{\pi 2^{\alpha}}{1-\alpha^2}\,\frac{c}{n^{\alpha}}\,.$$

*Proof.* Since $|f(x+u) - f(x)| \leqslant c|u|^{\alpha}$ we have, by (5.30) and (5.31),

$$|f(x) - \sigma_{n-1}(x)| \leqslant \frac{1}{\pi} \int_0^{\pi} 2cu^{\alpha}\,K_{n-1}(u)\,du$$

$$= \frac{2c}{\pi} \int_0^{\pi} \frac{2}{4n}\,\frac{u^{\alpha}\,\sin^2\,(nu/2)}{\sin^2\,(u/2)}\,du$$

$$\leqslant \frac{2^{\alpha+1}c}{n\pi} \int_0^{\pi/2} \frac{\pi^2}{4}\,\frac{\sin^2 nt}{t^{2-\alpha}}\,dt$$

where we have used the inequality $(\sin t)/t \geqslant 2/n$ for $0 < t \leqslant \frac{1}{2}\pi$. Hence

$$\|f - \sigma_{n-1}\|_c \leqslant \frac{c\,\pi}{n^\alpha\,2^{1-\alpha}} \int_0^{n\pi/2} \frac{\sin^2\xi}{\xi^{2-\alpha}}\,d\xi .$$

The last integral is less than or equal to

$$\int_0^1 \xi^\alpha\,d\xi + \int_1^\infty \frac{d\xi}{\xi^{2-\alpha}} = \frac{2}{1-\alpha^2} .$$

**5.43  EXAMPLE.** If $f \in C_{2\pi}$, that is $f \in C[-\pi,\pi]$ and is periodic, then

(a) $\qquad \sigma_{n-1}(x) = \frac{1}{\pi} \int_{-\infty}^\infty f\left(x + \frac{2t}{n}\right)\left(\frac{\sin t}{t}\right)^2 dt$ and

(b) $\qquad \|f(x) - \sigma_{n-1}(x)\|_c \leqslant \frac{2}{\pi}\,\omega\!\left(\frac{2}{n}\right) \{2 + \omega(\pi) + |\log \omega\!\left(\frac{2}{n}\right)|\}$

(here $\omega(\delta,f) = \omega(\delta) = \sup_{|x_1-x_2|\leqslant\delta} |f(x_1) - f(x_2)|$ is the modulus of continuity of $f$).

*Proof.*  From (5.30) we have

$$\sigma_{n-1}(x) = \frac{1}{n\pi} \int_0^\pi f(x+2\xi)\,\frac{\sin^2 n\xi}{\sin^2\xi}\,d\xi$$

$$= \frac{1}{n\pi} \int_0^\pi f(x+2\xi)\,\sin^2 n\xi \left(\sum_{k=-\infty}^\infty \frac{1}{(\xi+k\pi)^2}\right) d\xi .$$

By the Lebesgue convergence theorem

$$\sigma_{n-1}(x) = \lim_{m\to\infty} \frac{1}{n\pi} \int_0^\pi f(x+2\xi)\,\sin^2 n\xi \left(\sum_{-m}^m \frac{1}{(\xi+k\pi)^2}\right) d\xi$$

$$= \lim_{m\to\infty} \frac{1}{n\pi} \sum_{k=-m}^m \int_{k\pi}^{(k+1)\pi} f(x+2t)\,\frac{\sin^2 nt}{t^2}\,dt$$

$$= \frac{1}{n\pi} \int_{-\infty}^\infty f(x+2t)\,\frac{\sin^2 nt}{t^2}\,dt$$

$$= \frac{1}{\pi} \int_{-\infty}^\infty f\!\left(x+\frac{2t}{n}\right)\,\frac{\sin^2 t}{t^2}\,dt$$

When $f(x) = 1$, this gives

$$1 = \frac{1}{\pi} \int_{-\infty}^{\infty} \frac{\sin^2 t}{t^2} \, dt.$$

Hence

$$|f(x) - \sigma_{n-1}(x)| \leqslant \frac{1}{\pi} \int_{-\infty}^{\infty} |f\left(x + \frac{2t}{n}\right) - f(x)| \frac{\sin^2 t}{t^2} \, dt$$

$$\leqslant \frac{2}{\pi} \int_0^{\infty} \omega\left(\frac{2t}{n}\right) \frac{\sin^2 t}{t^2} \, dt$$

$$\leqslant \frac{2}{\pi} \{\omega\left(\frac{2}{n}\right) \int_0^1 \frac{\sin^2 t}{t^2} \, dt + \int_1^X \omega\left(\frac{2t}{n}\right) \frac{dt}{t^2}$$

$$+ \omega(\pi) \int_X^{\infty} \frac{dt}{t^2} \}$$

$$\leqslant \frac{2}{\pi} \{\omega\left(\frac{2}{n}\right) + \omega\left(\frac{2}{n}\right) \int_1^X \frac{(t+1) \, dt}{t^2} + \frac{\omega(\pi)}{X} \}.$$

Choose $X = 1/\omega(2/n)$ and (b) follows.

Remark. If $f$ also belongs to $\text{Lip}_c 1$ then (b) gives that

$$||f(x) - \sigma_{n-1}(x)||_c = O\left(\frac{\log n}{n}\right).$$

This estimate cannot be improved so far as the order of the right side is concerned (see N. Bary, *A Treatise on Trigonometric Series*, Volume 1, page 206; G. Meinardus, *Approximation of Functions: Theory and Numerical Methods*, page 49).

5.44 EXAMPLE. If a $2\pi$-periodic function $f(x)$ is of bounded variation over $[-\pi,\pi]$ then

$$a_n = O\left(\frac{1}{n}\right), \qquad b_n = O\left(\frac{1}{n}\right).$$

Let $V$ be the total variation over $[-\pi,\pi]$. Then

$$a_n = \frac{1}{\pi} \int_{-\pi}^{\pi} f(x) \cos nx \, dx = -\frac{1}{\pi} \int_{-\pi}^{\pi} f\left(x + \frac{\pi}{n}\right) \cos nx \, dx.$$

Hence

$$|a_n| \leqslant \frac{1}{2\pi} \int_{-\pi}^{\pi} |f\left(x + \frac{\pi}{n}\right) - f(x)| \, dx.$$

By periodicity, we have

$$|a_n| \leqslant \frac{1}{2\pi} \int_{-\pi}^{\pi} | f\left(x + \frac{k\pi}{n}\right) - f\left(x + (k-1)\frac{\pi}{n}\right) | \, dx.$$

Adding such inequalities for $k = 1, 2, \ldots, 2n$ we have

$$2n\,|a_n| \leqslant \frac{1}{2\pi} \int_{-\pi}^{\pi} \sum_{k=1}^{2n} | f\left(x + \frac{k\pi}{n}\right) - f\left(x + (k-1)\frac{\pi}{n}\right) | \, dx$$

$$\leqslant V, \qquad n \geqslant 1.$$

Similarly $2n\,|b_n| \leqslant V$, $\quad n \geqslant 1$.

*Remark:* By Theorem 5.37(i) the Fourier series is $(C,1)$ summable at $x$ to $\frac{1}{2}\{f(x+0) + f(x-0)\}$ and hence by Theorem 4.16, it is convergent at every point $x$ to $\frac{1}{2}\{f(x+0) + f(x-0)\}$.

**5.45  EXAMPLE.** If $f(x)$ is bounded and has Fourier coefficients $O(1/n)$ then the partial sums $S_n(x)$ are uniformly bounded.
By (5.30) and (5.27, iv)

$$|\sigma_n(x)| \leqslant M = \sup |f(x)|$$

and by (5.29)

$$|S_n(x) - \sigma_n(x)| = |\frac{1}{n+1} \sum_{k=1}^{n} k(a_k \cos kx + b_k \sin kx)|.$$

Let $U = \sup_{n \geqslant 1} \{\max(n\,|a_n|,\ n\,|b_n|)\}$. Then $|S_n(x)| < M + 2U$. If the Fourier coefficients satisfy the inequalities of Example 5.44 then $|S_n(x)| < M + V$.

**5.46  THEOREM.** (Fejér–Lebesgue)  *The Fourier series of $f$ is $(C,1)$ summable at $x$ to $f(x)$ for each $x$ such that*

$$\lim_{h \to 0} \frac{1}{h} \int_0^h |f(u+x) - f(x)| \, du = 0, \tag{5.33}$$

*in particular, it is summable $(C,1)$ to $f(x)$ for almost all $x$.*

*Proof.* Let $x$ be a point where (5.33) is satisfied and write

$$\phi(u) = f(x+u) + f(x-u) - 2f(x), \quad \Phi(t) = \int_0^t |\phi(u)| \, du.$$

Therefore

$$|\int_0^h |\phi(u)| \, du | \leqslant |\int_0^h |f(x+u) - f(x)| \, dx | + |\int_0^h |f(x-u) - f(x)| \, dx |.$$

Hence, given $\varepsilon > 0$, we can find $\eta$ $(0 < \eta \leqslant 1)$ such that

$$\Phi(h) \leqslant h\varepsilon \qquad \text{for } h \in (0, \eta). \tag{5.34}$$

Choose $n_0 > 1/\eta$ such that $\Phi(\pi) < \varepsilon\eta^2\, n$ for $n \geqslant n_0$. Then, for $n \geqslant n_0$, we have from (5.32) (with $s = f(x)$) that

$$|\sigma_n(x) - f(x)| \leqslant \frac{1}{\pi} [\int_0^{1/n} |\phi(u)| K_n(u)\,du + \int_{1/n}^{\eta} |\phi(u)| K_n(u)\,du$$

$$+ \int_{\eta}^{\pi} |\phi(u)| K_n(u)\,du\,]$$

$$= \frac{1}{\pi}(I_1 + I_2 + I_3) \text{ say.}$$

By (5.27,iii) and (5.34) we have $I_1 \leqslant \frac{1}{2}(n+1)\,\Phi(1/n) < \frac{1}{2}(n+1)\,(\varepsilon/n) \leqslant \varepsilon$. Using (5.27,v) and partial integration we obtain

$$I_2 < \int_{1/n}^{\eta} |\phi(u)| \frac{\pi^2}{2(n+1)u^2}\,du$$

$$= \frac{\pi^2}{2(n+1)} [\frac{\Phi(\eta)}{\eta^2} - n^2\Phi(\frac{1}{n}) + 2\int_{1/n}^{\eta} \frac{\Phi(u)}{u^3}\,du\,]$$

$$\leqslant \frac{\pi^2}{2n} [\frac{\varepsilon}{\eta} + 2\varepsilon \int_{1/n}^{\eta} \frac{du}{u^2}\,] < \pi^2\,\varepsilon.$$

By (5.27,v) we have

$$I_3 \leqslant \int_{\eta}^{\pi} |\phi(u)| \frac{\pi^2}{2(n+1)} \frac{du}{u^2} < \frac{\pi^2}{2(n+1)\eta^2}\,\varepsilon\eta^2 n < \frac{\pi^2\varepsilon}{2}.$$

Hence, for $n \geqslant n_0$,

$$\pi |\sigma_n(x) - f(x)| \leqslant (\tfrac{3}{2}\pi^2 + 1)\varepsilon. \quad \triangle$$

We note that condition (5.33) is satisfied for almost all values of x. The set of points x where (5.33) holds is called the Lebesgue set for $f$ (see I. P. Natanson, *Theory of Functions of a Real Variable*, Volume 1, pages 255-6). Thus, $\sigma_n(x)$ converges to $f(x)$ almost everywhere.

If $f \in C\,[-\pi, \pi]$ and is periodic then the convergence of $\sigma_n(x)$ to $f(x)$ is uniform on any interval $[A, B]$ (see Theorem 5.37(ii) and 5.37(iii)). But, if we consider the convergence of $S_n(x)$ then the situation is completely different. There exist continuous functions, $f$, such that the Fourier series of $f$ diverges on a dense set of points.

5.47  EXAMPLE. (Fejér) (A continuous function with a divergent Fourier series.)

Let

$$P(x,m,n) = \frac{\cos mx}{n} + \frac{\cos(m+1)x}{n-1} + \ldots + \frac{\cos(m+n-1)x}{1}$$

$$- \frac{\cos(m+n+1)}{1} - \frac{\cos(m+n+2)x}{2} - \ldots - \frac{\cos(m+2n)x}{n}.$$

Then

$$P(x,m,n) = \sum_{k=1}^{n} \frac{\cos(m+n-k)x - \cos(m+n+k)x}{k}$$

$$= \sum_{k=1}^{n} \frac{2\sin(m+n)x \, \sin kx}{k}$$

$$= 2\sin(m+n)x \left( \sum_{k=1}^{n} \frac{\sin kx}{k} \right).$$

Hence (see Example 5.5(ii))

(a)                          $|P(x,m,n)| \leqslant 4\sqrt{\pi}$.

Let $m_j = n_j = 2^{j^2}$ and let

(b)                          $f(x) = \sum_{k=1}^{\infty} \frac{P(x,m_k,n_k)}{k^2}$.

The inequality (a) shows that the right-hand series of (b) is absolutely and uniformly convergent on any interval $[A,B]$. Since $P(x,m_k,n_k)$ are trigonometric polynomials, $f(x) \in C_{2\pi}$. By uniform convergence we have

$$b_p(f) = \frac{1}{\pi} \int_{-\pi}^{\pi} f(x) \sin px \, dx = \frac{1}{\pi} \sum_{k=1}^{\infty} \int_{-\pi}^{\pi} \frac{P(x,m_k,n_k)}{k^2} \sin px \, dx.$$

Since $\int_{-\pi}^{\pi} \cos qx \sin px \, dx = 0$ ($p,q$ any integers or zero)

$$b_p(f) = 0.$$

Similarly

$$a_0(f) = 0.$$

Also, for $p \geqslant 1$,

$$a_p(f) = \frac{1}{\pi} \int_{-\pi}^{\pi} f(x) \cos px \, dx$$

$$= \frac{1}{\pi} \sum_{1}^{\infty} \int_{-\pi}^{\pi} \frac{P(x, m_k, n_k)}{k^2} \cos px \, dx.$$

Note that $m_k + 2n_k < m_{k+1}$ and so if $\cos px$ occurs in the group of terms $P(x, m_k, n_k)$ (for some $k$) then it occurs once only and it cannot occur in the group of terms $P(x, m_j, n_j)$, $j \neq k$    (thus if $p = 6$ then $\cos 6x$ occurs in the group $P(x, m_1, n_1)$ and

$$a_6(f) = \frac{1}{\pi} \int_{-\pi}^{\pi} \frac{P(x, m_1, n_1)}{1^2} \cos 6x \, dx$$

$$= \frac{1}{\pi} \frac{1}{1^2} \int_{-\pi}^{\pi} \frac{(-\cos 6x) \cos 6x}{2} \, dx$$

$$= \frac{-1}{1^2} \frac{1}{2} \; ;$$

further

$$a_7 = \ldots = a_{15} = 0, \qquad a_{16} \neq 0).$$

It follows from this that if $\sum_{k=1}^{\infty} \frac{1}{k^2} P(x, m_k, n_k)$ is written as

$$\sum_{p=1}^{\infty} a_p(f) \cos px$$

then this series is the Fourier series of $f(x)$.    Let $S(n, x)$ be the sum of the first $n$ terms of this series.

Then

$$S(2n_1 + 2n_2 + \ldots + 2n_{k-1} + n_k, \; 0)$$

$$-S(2n_1 + 2n_2 + \ldots + 2n_{k-1} \qquad , \; 0)$$

$$= \frac{1}{k^2} \left( \frac{1}{n_k} + \frac{1}{n_k - 1} + \ldots + 1 \right) > \frac{\log n_k}{k^2} = \log 2.$$

This shows that $S(n, 0)$ does not tend to a limit. Thus the Fourier series of the continuous function $f(x)$ diverges for $x = 0$.

By considering the function

$$F(x) = c_1 f(x-r_1) + c_2 f(x-r_2) + \ldots$$

where $r_1$, $r_2$, ... is an everywhere dense set of rational numbers on $0 \leqslant x \leqslant 2\pi$ and $\sum_{k=1}^{\infty} c_k$ is a convergent series of positive terms, it can be shown that $F(x)$ is continuous and the Fourier series of $F(x)$ diverges at each of the points of the set $r_1$, $r_2$, .... We omit the details (see A. Zygmund, *Trigonometric Series*, Vol. 1, pages 298–301; R. L. Jeffrey, *Trigonometric Series*, pages 20–3).

A recent result due to Kahane and Katznelson ('Sur les ensembles de divergence des séries trigonometriques', *Studia Math.* **26** (1966), pages 305–6) is as follows: Given a set $E$ of measure zero, there exists a continuous function whose Fourier series diverges at every point of $E$. This gives the utmost information for if $f$ is continuous then $f \in L^2$ and, so, $S_n(x)$ converges almost everywhere to $f(x)$ (see Lennart Carleson, *Acta Math. Stockh.* **116** (1966), pages 135–57).

5.48  THEOREM. *The trigonometric systems* (5.13) *and* (5.14) *are complete in the space* $L[-\pi, \pi)$.

*Proof.*  We have to show that if $f \in L[-\pi, \pi)$ and if

$$\int_{-\pi}^{\pi} f(x) \cos kx \, dx = 0, \qquad k = 0, 1, \ldots \tag{5.35}$$

and

$$\int_{-\pi}^{\pi} f(x) \sin kx \, dx = 0, \qquad k = 1, 2, \ldots \tag{5.36}$$

then $f(x) = 0$ almost everywhere in $[-\pi, \pi)$.

First, suppose that $f(x) \in C[-\pi, \pi]$. Then (5.35) and (5.36) imply that the Fourier coefficients of $f$ are all zero and, so, $\sigma_n(x) \equiv \sigma_n(x, f)$ vanishes identically (see (5.29)). By Theorem 5.37 we have that $f(x) = \lim_{n \to \infty} \sigma_n(x) = 0$ for $-\pi < x < \pi$. Therefore, by continuity $f(x) = 0$ for $-\pi \leqslant x \leqslant \pi$.

Suppose, now, that $f(x) \in L[-\pi, \pi)$. Consider the function $F$ where

$$F(x) = \int_{-\pi}^{x} f(t) \, dt \text{ for } -\pi \leqslant x < \pi \text{ and } F(x+2\pi) = F(x) \text{ outside } [-\pi, \pi).$$

$$\tag{5.37}$$

Therefore, $F(\pi) = F(-\pi) = 0$. Since $a_0 = \dfrac{1}{\pi} \int_{-\pi}^{\pi} f(t) \, dt = 0$ we have that $F(x)$ is continuous not only on $[-\pi, \pi]$ but on the entire line $-\infty < x < \infty$. When $n > 0$, integration by parts yields

$$a_n(F) = \frac{1}{\pi} \int_{-\pi}^{\pi} F(t) \cos nt \, dt = \frac{-1}{\pi n} \int_{-\pi}^{\pi} f(t) \sin nt \, dt = 0.$$

Similarly, $b_n(F) = 0$. Hence, $G(x) = F(x) - (a_0(F))/2$ is continuous and has all its Fourier coefficients equal to zero. Therefore, by the first case, $G(x) \equiv 0$ and, thus, $F(x) \equiv a_0(F)/2$. From (5.37) we see that $F^{(1)}(x) = f(x)$ almost everywhere and, so, $f(x) = 0$ almost everywhere in $[-\pi, \pi)$.   $\triangle$

5.49   COROLLARY.   If two functions $g$, $G \in L$ $[-\pi, \pi)$ have the same Fourier series, i.e. $a_n(g) = a_n(G)$ $(n = 0, 1, \ldots)$ and $b_n(g) = b_n(G)$ $(n = 1, 2, \ldots)$, then $g(x) = G(x)$ almost everywhere on $[-\pi, \pi)$.

Proof.   Take $f(x) = g(x) - G(x)$ in Theorem 5.48.   $\triangle$

An alternative proof which appeals to Theorem 5.46 is as follows. By hypothesis we have $\sigma_n(x, g) = \sigma_n(x, G)$. Since $\sigma_n(x, g) \to g(x)$ almost everywhere and $\sigma_n(x, G) = G(x)$ almost everywhere we have $g(x) = G(x)$ almost everywhere.

5.50   THEOREM. (Parseval)   If $f$, $F \in L^2$ $[-\pi, \pi)$, are real valued and

$$f(x) \simeq \frac{a_0}{2} + \sum_{n=1}^{\infty} (a_n \cos nx + b_n \sin nx) \tag{5.38}$$

and

$$F(x) \simeq \frac{\alpha_0}{2} + \sum_{n=1}^{\infty} (\alpha_n \cos nx + \beta_n \sin nx)$$

then

$$\frac{1}{\pi} \int_{-\pi}^{\pi} (f(x))^2 \, dx = \frac{1}{2} a_0^2 + \sum_{n=1}^{\infty} (a_n^2 + b_n^2) \tag{5.39}$$

and

$$\frac{1}{\pi} \int_{-\pi}^{\pi} f(x) F(x) \, dx = \frac{1}{2} a_0 \alpha_0 + \sum_{n=1}^{\infty} (a_n \alpha_n + b_n \beta_n). \tag{5.40}$$

Proof.   Since $f \in L^2$ $[-\pi, \pi)$, Bessel's inequality gives the convergence of the series on the right-hand side of (5.39). By the Riesz-Fischer theorem there exists a function $g \in L^2$ $[-\pi, \pi)$ such that the series on

the right-hand side of (5.38) is its Fourier series. Since this series is also the Fourier series of $f$, Corollary 5.49 gives us that $f(x) = g(x)$ almost everwhere. By (5.18) and the triangle inequality we have
$$| \; ||f|| - ||S_n(f)|| \; | \leqslant ||f - S_n(f)|| = o(1). \quad \text{But,}$$

$$||S_n(f)||^2 = (S_n(f), S_n(f))$$

$$= \pi \left( \frac{a_0^2}{2} + \sum_{k=1}^{n} (a_k^2 + b_k^2) \right).$$

Letting $n \to \infty$ we have

$$\pi \left( \frac{a_0^2}{2} + \sum_{k=1}^{\infty} (a_k^2 + b_k^2) \right) = ||f||^2$$

and (5.39) is proved.

To prove (5.40) consider Parseval's formula (5.39) for $f + F$ and $f - F$, and subtract the two. We obtain

$$\frac{1}{\pi} \int_{-\pi}^{\pi} [ (f(x) + F(x))^2 - (f(x) - F(x))^2 ] \, dx$$

$$= \frac{1}{2} [ (a_0 + \alpha_0)^2 - (a_0 - \alpha_0)^2 ]$$

$$+ \sum_{n=1}^{\infty} [ (a_n + \alpha_n)^2 - (a_n - \alpha_n)^2 + (b_n + \beta_n)^2 - (b_n - \beta_n)^2 ]. \quad \triangle$$

*Exercise.* (i) Prove that if $f \in L^2 [-\pi, \pi)$ and $f(x) \simeq \sum_{n=-\infty}^{\infty} c_n e^{inx}$ then $\dfrac{1}{2\pi} \int_{-\pi}^{\pi} |f(x)|^2 \, dx = \sum_{n=-\infty}^{\infty} |c_n|^2$.

(ii) Suppose that for the trigonometric series (5.1) with $a_k$, $b_k$ real, $|\sigma_n(x)| \leqslant M$ for all $n$ and $x$. Prove that (5.1) is the Fourier series of a function $f \in L^2$. (*Hint*: From (5.29) we obtain

$$\frac{1}{\pi} \int_{-\pi}^{\pi} (\sigma_n(x))^2 \, dx = \frac{1}{2} a_0^2 + \sum_{k=1}^{n} (a_k^2 + b_k^2) (1 - \frac{k}{n+1})^2.$$

Hence, for $0 < m \leqslant n$,

$$2M^2 \geqslant \frac{1}{2} a_0^2 + \sum_{k=1}^{n} (a_k^2 + b_k^2) ( 1 - \frac{k}{n+1})^2.$$

Now, let $n \to \infty$ and apply Theorem 5.21.)

(iii) Suppose that $a_k$ and $b_k$ are real in the trigonometric series (5.1). Prove that if $\sigma_n(x)$ converges uniformly on $[-\pi, \pi]$ then (5.1) is the Fourier series of a continuous and periodic function $f$.    (*Hint*: By periodicity $\{\sigma_n(x)\}_0^\infty$ converges uniformly on any interval $[A,B]$. Let $f(x) = \lim_{n\to\infty} \sigma_n(x)$.    Thus,

$$\frac{1}{\pi} \int_{-\pi}^{\pi} \sigma_n(x) \cos kx \, dx = (1 - \frac{k}{n+1}) a_k$$

for $0 \leqslant k \leqslant n$.   Fix $k$ and let $n\to\infty$.   The case is similar for $b_k$.)

   (iv)  Prove that a necessary and sufficient condition for the Fourier series of $f$ to be $(C,1)$ summable at a point $x$ to the value $s$ is that $\lim_{n\to\infty} \int_0^\delta [f(x+u) + f(x-u) - 2s] K_n(u) \, du = 0$.    (*Hint*: Use (5.32) and (5.27,v).)

   (v)   Let $f \in C[a,b]$, $[a',b'] \subseteq (a,b)$, and $0 < \delta < \min(\pi, b-b', a'-a)$.    Prove that for the uniform convergence of the Fourier series of $f$ on $[a',b']$ it is necessary and sufficient that

$$\lim_{n\to\infty} \int_0^\delta [f(x+u) + f(x-u) - 2f(x)] \frac{\sin nu}{u} \, du = 0$$

uniformly on $[a',b']$.

   (vi)  Prove that any Fourier series may be integrated term by term.   Deduce that if $a_n$ and $b_n$ are the Fourier coefficients of a real valued function then $\sum_{n=1}^{\infty} b_n/n$ is convergent.        (*Hint*: Consider $F(x) = \int_{-\pi}^{x} (f(t) - \frac{1}{2} a_0) \, dt$ and observe that $F(-\pi) = F(\pi) = 0$.)

   (vii) Prove that the series $\sum_{n=2}^{\infty} \sin nx/\log n$ is convergent for all values of $x$ but it is not a Fourier series.   (*Hint*: See Exercise (i), Chapter 1).   It cannot be a Fourier series since

$$\sum_{n=2}^{\infty} \frac{b_n}{n} = \sum_{n=2}^{\infty} \frac{1}{n \log n} = \infty .)$$

   (viii) Suppose $\lambda_n$ is positive and decreases to $0$ and that $n\lambda_n = O(1)$.   Prove that the series

$$\sum_1^\infty \lambda_n \sin nx$$

converges boundedly on $[-\pi,\pi]$ to a function of which it is the Fourier series.   (*Hint*: See Example 5.5(ii).   The series converges for all $x$. Let its sum be $f(x)$.   Since $\sin nx$ is odd we need consider the interval $[0,\pi]$.   Let $\sup_{m \geqslant n} m\lambda_m = \mu_n \leqslant \mu$.

Then for $x \geqslant \pi/k$

$$\left| \sum_{n=k}^{m} \sin nx \right| \leqslant \frac{1}{\sin (x/2)} \leqslant \frac{\pi}{x}$$

and Abel's lemma gives

$$\left| \sum_{n=k}^{m} \lambda_n \sin nx \right| \leqslant k\lambda_k \leqslant \mu_k.$$

If $x \leqslant \pi/m$ then

$$\left| \sum_{n=k}^{m} \lambda_n \sin nx \right| \leqslant x \sum_{n=k}^{m} n\lambda_n \leqslant \frac{\pi}{m} (\mu_k)m = \pi\mu_k.$$

If $\pi/m < x < \pi/k$ then suppose $x \in [\pi/(p+1), \pi/p]$ and combine the above two arguments. We have that

$$\left| \sum_{n=k}^{m} \lambda_n \sin nx \right| \leqslant \left| \sum_{n=k}^{p} \right| + \left| \sum_{m=p+1}^{m} \right| \leqslant xp\mu_k + \lambda_{p+1} \frac{\pi}{x} \leqslant \pi\mu_k + \mu_k.$$

Now note that

$$\int_0^{\pi} |f| \, dx = \sum_1^{\infty} \int_{\pi/k+1}^{\pi/k} |f| \, dx \leqslant (\pi+1)\mu \sum_{k=1}^{\infty} \frac{\pi}{k(k+1)} .)$$

(ix) Let $\lambda_n$ be as in Exercise (viii) and suppose $n\lambda_n = o(1)$. Prove that $\sum_1^{\infty} \lambda_n \sin nx$ converges uniformly to a continuous function of which it is the Fourier series.

### Abel-Poisson Summability of Fourier Series

We recall that the series $\sum_{n=0}^{\infty} A_n(x)$, where $A_n(x) = a_n \cos nx + b_n \sin nx$, $A_0 = a_0/2$, is said to be summable by the Abel-Poisson method, or briefly, summable $A$ at a point $x_0$ to the value $s = s(x_0)$ if the series $\sum_{n=0}^{\infty} A_n(x_0) r^n$ converges for $0 < r < 1$ and

$$\lim_{r \to 1-} \left( \sum_{n=0}^{\infty} A_n(x_0) r^n \right) = s.$$

Since we know that a series summable $(C,1)$ is summable $A$ to the same sum (see Chapter 1), it follows from Theorems 5.37(i) and 5.46 that the Fourier series of $f$ is summable $A$ at the point $x$ to $f(x)$ for almost all values of $x$ and it is summable $A$ at the point $x$ to the sum

$\frac{1}{2}[f(x+0) + f(x-0)]$ whenever this expression has a meaning. We assume, as before, that $f \in L\ [-\pi,\pi)$ and is periodic.

We shall now obtain an integral expression for the so-called Poisson sum

$$f(r,x) = \sum_{n=0}^{\infty} A_n(x) r^n \qquad (0 < r < 1) \qquad (5.41)$$

of the series $\sum_{n=0}^{\infty} A_n(x)$. Since $A_n(x) = o(1)$, the series on the right of (5.41) is absolutely and uniformly convergent for $0 \leqslant r \leqslant 1-\delta,\ \delta > 0$, and we have

$$f(r,x) = \frac{1}{2\pi} \int_{-\pi}^{\pi} f(t)\ dt + \frac{1}{\pi} \sum_{n=1}^{\infty} r^n \int_{-\pi}^{\pi} f(t) \cos n(t-x)\ dt.$$

By Lebesgue's dominated convergence theorem,

$$f(r,x) = \frac{1}{\pi} \int_{-\pi}^{\pi} f(t) \{ \frac{1}{2} + \sum_{n=1}^{\infty} r^n \cos n(t-x) \}\ dt. \qquad (5.42)$$

Now

$$\frac{1}{2} + \sum_{n=1}^{\infty} r^n \cos n\phi = \mathrm{Re}\{ \frac{1}{2} + \sum_{n=1}^{\infty} r^n e^{in\phi} \}$$

$$= \mathrm{Re}\{ \frac{1}{2} + \frac{re^{i\phi}}{1-re^{i\phi}} \} = \mathrm{Re}\{ \frac{1+re^{i\phi}}{2(1-re^{i\phi})} \}$$

$$= \frac{1-r^2}{2(1-2r \cos \phi + r^2)} = P(r,\phi) \text{ say.} \quad (5.43)$$

From (5.42) and (5.43) we have, for $0 \leqslant r < 1$,

$$f(r,x) = \frac{1}{2\pi} \int_{-\pi}^{\pi} f(t)\ \frac{1-r^2}{1-2r \cos (t-x) + r^2}\ dt. \qquad (5.44)$$

The expression $P(r,\phi)$ is known as a Poisson kernel and the right-hand side of (5.44) is known as the Poisson integral of $f$. If we take $f(x) = 1$, then $A_0 = 1$, $A_n = 0$ $(n \geqslant 1)$, $f(r,x) = 1$ and we obtain

$$1 = \frac{1}{\pi} \int_{-\pi}^{\pi} P(r,t-x)\ dt. \qquad (5.45)$$

5.51 THEOREM. *If f is real valued and* $m \leqslant f(x) \leqslant M$ *for all* $x$, *then* $m \leqslant f(r,x) \leqslant M$ $(0 \leqslant r \leqslant 1,\ -\pi \leqslant x \leqslant \pi)$.

*Proof.*   From (5.44) and (5.45) we have

$$f(r,x)-m \;=\; \frac{1}{\pi}\int_{-\pi}^{\pi}(f(t)-m)\,P(r,t-x)\,dt.$$

Now $1-2r\cos\phi+r^2=(1-r)^2+4r\sin^2(\phi/2)>0$, and so $P(r,t-x)>0$ and we have $f(r,x)\geqslant m$.

The remaining part of the theorem can be similarly proved by considering $M-f(r,x)$.   $\triangle$

*Exercise.*   (i)   Prove that if $0\leqslant r<1$

then    $\mathrm{Im}\left(\dfrac{1+re^{i\phi}}{2(1-re^{i\phi})}\right) = \dfrac{r\sin\phi}{1-2r\cos\phi+r^2} = \displaystyle\sum_{n=1}^{\infty} r^n\sin n\phi.$

Hence, show that

$$\sum_{n=1}^{\infty}(a_n\sin nx-b_n\cos nx)r^n = \frac{-1}{\pi}\int_{-\pi}^{\pi}f(t)\left(\frac{r\sin(t-x)}{1-2r\cos(t-x)+r^2}\right)dt$$

(ii)   Let $a_n,\ b_n$ be the Fourier coefficients of a real valued function of $f$.   Prove that the series

$$\phi(z) = \tfrac{1}{2}a_0 + \sum_{n=1}^{\infty}(a_n-ib_n)\,z^n$$

is convergent for $|z|<1$.   If $z=re^{i\theta}$, and $r<1$ show that

$$\mathrm{Re}\,(\phi(re^{i\theta})) = \sum_{n=0}^{\infty} A_n r^n = f(r,\theta)$$

and

$$\mathrm{Im}\,(\phi(re^{i\theta})) = \sum_{n=1}^{\infty}(a_n\sin n\theta-b_n\cos n\theta)\,r^n.$$

Deduce that $f(r,\theta)$ is a harmonic function for $r<1$.   (Note that $a_n=o(1)$, $b_n=o(1)$ by Riemann-Lebesgue theorem.)

(iii)   Prove the following properties of a Poisson kernel:

(a)   If $|\phi|\geqslant\delta>0$, then $m(r,\delta)=\max_{\delta\leqslant|\phi|\leqslant\pi}P(r,\phi)\to 0$ as $r\to 1-$.

(b)   $\lim_{r\to 1-}\dfrac{2}{\pi}\int_{0}^{\delta}P(r,\phi)\,d\phi = 1.$

*(Hint:*   (a)   $1-2r\cos\phi+r^2\geqslant 1-2r\cos\delta+r^2$ for $\delta\leqslant|\phi|\leqslant\pi$.

(b)   From (5.45) (with $x=0$) we have that

$$1=\frac{1}{\pi}\int_{-\pi}^{\pi}P(r,\phi)\,d\phi=\frac{2}{\pi}\int_{0}^{\delta}P(r,\phi)\,d\phi+\frac{2}{\pi}\int_{\delta}^{\pi}P(r,\phi)\,d\phi.)$$

We now have, for almost all values of $x$, that

$$\lim_{r \to 1-} f(r,x) = f(x).$$

We now consider a non-radial approach of the point $(r,x)$ to a point $(1,x_0)$ on the unit circumference.

5.52  THEOREM. (Fatou)  If $f(x)$ is continuous at $x = x_0$, then

$$f(r,x) = \frac{1}{\pi} \int_{-\pi}^{\pi} f(t) \, P(r,t-x) \, dt$$

tends to $f(x_0)$ as the point $(r,x)$ tends to the point $(1,x_0)$ along a path which lies always in the unit disc.

Proof.  From (5.44) and (5.45) we have

$$f(r,x) - f(x_0) = \frac{1}{\pi} \int_{-\pi}^{\pi} (f(t) - f(x_0)) \, P(r,t-x) \, dt.$$

Since $f(x)$ is continuous at $(x_0)$ we have

$$|f(t) - f(x_0)| < \varepsilon \quad \text{for } |t-x_0| < \delta.$$

Hence

$$|f(r,x) - f(x_0)| \leqslant \frac{1}{\pi} \{ \int_{-\pi}^{x_0-\delta} + \int_{x_0-\delta}^{x_0+\delta} + \int_{x_0+\delta}^{\pi} \} \, |f(t) - f(x_0)| \, P(r,t-x) \, dt$$

$$= I_1 + I_2 + I_3 \quad \text{say.}$$

Then

$$I_2 < \frac{\varepsilon}{\pi} \int_{x_0-\delta}^{x_0+\delta} P(r,t-x) \, dt < \frac{\varepsilon}{\pi} \int_{-\pi}^{\pi} P(r,t-x) \, dt = \varepsilon.$$

For $I_1$ and $I_3$ we have (see Exercise (iii) following Theorem 5.51)

$$I_j < \frac{\varepsilon}{\pi} \int_{-\pi}^{\pi} (|f(t)| + |f(x_0)|) \, dt \quad j = 1, 3,$$

for $r_0(\varepsilon) < r < 1$.  Hence the theorem is proved.    $\triangle$

Remark.  Let $f \in C_{2\pi}$ and be real valued.  Then

$$f(r,x) = \text{Re} \left( \frac{1}{2\pi} \int_{-\pi}^{\pi} f(t) \frac{e^{it}+z}{e^{it}-z} \, dt \right), \quad z = re^{ix}, \ r < 1,$$

is a harmonic function in the disc $D = \{z: |z| < 1\}$. Furthermore, if the complex coordinate of any point $P$ on the unit circumference $|z| = 1$ is denoted by $\xi = e^{ix_0}$, then $\lim_{z \to \xi, |z| < 1} f(r, x) = f(x_0)$.

Hence Poisson's integral (5.46) (where $f(t)$ is now real-valued) solves the Dirichlet problem for the unit disc: given $f \in C_{2\pi}$ and real-valued, to find a function $f(r, x)$ harmonic in $D$ and such that $f(r, x)$ tends to the limit $f(x_0)$ when $z = re^{ix}$ approaches, from inside the unit disc, $e^{ix_0}$.

*Exercise.* Show that if $f(x_0 + 0)$ and $f(x_0 - 0)$ exist then

$$\lim_{r \to 1-} f(r, x_0) = \frac{f(x_0 + 0) + f(x_0 - 0)}{2}.$$

### Riemann's Method of Summation

We now define Riemann's method of summation and prove that it is regular.

5.53 DEFINITION. *A series* $\sum_{n=0}^{\infty} A_n$ *is summable by Riemann's method (or, briefly, summable R) to the value* $S$ *if the series* $\sum_{n=0}^{\infty} R_n(h)$, $R_0(h) = A_0$, $R_n(h) = A_n ((\sin nh)/nh)^2$ $(n \geqslant 1)$, *is convergent for every* $h \neq 0$ *and*

$$\lim_{h \to 0^+} \sum_{n=0}^{\infty} R_n(h) = S.$$

5.54 THEOREM. *Let the series* $\sum_{n=0}^{\infty} A_n$ *be convergent and* $S$ *be its sum. The series* $\sum_{n=0}^{\infty} R_n(h)$ *is convergent for every* $h \neq 0$ *and* $\lim_{h \to 0^+} \sum_{n=0}^{\infty} R_n(h) = S.$

*Proof.* Since $\sum_{n=0}^{\infty} A_n$ is convergent we have that $A_n = o(1)$ and $R_n(h) = O(1/n^2 h^2)$. Hence $R(h) \equiv \sum_{n=0}^{\infty} R_n(h)$ is convergent for $h \neq 0$.

Write $r_n = \sum_{k=n+1}^{\infty} A_k$. Let $\varepsilon > 0$ be given. There exists $N = N(\varepsilon)$ such that

$$|r_n| < \varepsilon \quad \text{for } n \geqslant N.$$

Now,

$$R(h) = (A_0 + \sum_{n=1}^{N} A_n \left(\frac{\sin nh}{nh}\right)^2) + \sum_{n=N+1}^{\infty} A_n \left(\frac{\sin nh}{nh}\right)^2$$

$$= \Sigma_1 + \Sigma_2 \quad \text{say.} \tag{5.47}$$

We fix $N$. There exists $h_0 = h_0(\varepsilon)$ such that for $0 < h < h_0$ we have

$$\left| A_0 + \sum_{n=1}^{N} A_n \left(\frac{\sin nh}{nh}\right)^2 - \sum_{n=0}^{N} A_n \right| < \varepsilon.$$

Hence, from (5.47) and (5.48), we have, for $0 < h < h_0$,

$$|\Sigma_1 - S| = \left| A_0 + \sum_{n=1}^{N} A_n \left(\frac{\sin nh}{nh}\right)^2 - \sum_{n=0}^{N} A_n - \sum_{n=N+1}^{\infty} A_n \right| < 2\varepsilon.$$

Also

$$\Sigma_2 = \sum_{n=N+1}^{\infty} (r_{n-1} - r_n) \left(\frac{\sin nh}{nh}\right)^2$$

$$= \lim_{k \to \infty} \left\{ \sum_{n=N+1}^{k} (r_{n-1} - r_n) \left(\frac{\sin nh}{nh}\right)^2 \right\}$$

$$= \lim_{k \to \infty} \left[ \sum_{n=N+1}^{k} \left\{ r_n \left( \left(\frac{\sin (n+1) h}{(n+1) h}\right)^2 - \left(\frac{\sin nh}{nh}\right)^2 \right) \right\} \right.$$

$$\left. + r_n \left(\frac{\sin (N+1) h}{(N+1) h}\right)^2 - r_k \left(\frac{\sin (k+1) h}{(k+1) h}\right)^2 \right].$$

Now the last term tends to zero as $k \to \infty$ since $r_k = o(1)$.   Hence

$$|\Sigma_2| \leqslant \varepsilon \sum_{n=N+1}^{\infty} \left| \left(\frac{\sin (n+1) h}{(n+1) h}\right)^2 - \left(\frac{\sin nh}{nh}\right)^2 \right| + \varepsilon$$

$$= \sum_{n=N+1}^{\infty} \left| \int_{nh}^{(n+1)h} \frac{d}{dt} \left(\frac{\sin t}{t}\right)^2 dt \right| + \varepsilon$$

$$\leqslant \varepsilon + \varepsilon \int_{(N+1)h}^{\infty} \left| \frac{d}{dt} \left(\frac{\sin t}{t}\right)^2 \right| dt$$

$$\leqslant \varepsilon + 2\varepsilon \int_{0}^{\infty} \left| \left(\frac{\sin t}{t}\right) \left(\frac{t \cos t - \sin t}{t^2}\right) \right| dt = \varepsilon + 2\varepsilon I \quad \text{say.}$$

Since $t \cos t - \sin t = O(t^3)$ as $t \to 0$, the integrand in $I$ is bounded near

$t = 0$. Also it is $O(1/t^2)$ as $t-\infty$.. Hence the integral $I$ is convergent and we have

$$|\Sigma_1 + \Sigma_2 - S| < 3\varepsilon + 2\varepsilon I,$$

hence, the theorem follows. $\triangle$

The Fourier series of a function $f \in L[-\pi, \pi)$ is summable $R$ almost everywhere to this function. Hence the Riemann method of summation is Fourier-effective. For a trigonometric series the following important statement on the uniqueness of the expansion of a function into a trigonometric series can be proved by an application of this summation method: If a trigonometric series with coefficients tending to zero is summable to zero by Riemann's method everywhere apart, perhaps, from a finite number of points, then all its coefficients equal zero.

For the proofs of these and extensions we refer the reader to N. Bary, *A Treatise on Trigonometric Series*, Volume 1, pages 189–95 and A. Zygmund, *Trigonometric Series*, Volume 1, Chapter IX.

### Absolute Convergence

If the series

$$\sum_{n=1}^{\infty} (|a_n| + |b_n|) \tag{5.49}$$

is convergent, then the trigonometric series (5.1) is absolutely and uniformly convergent over any interval $[A, B]$. It is then a Fourier series of a function $f \in C_{2\pi}$. We now prove a theorem which gives the convergence of the series (5.49).

5.55 THEOREM. (Lusin–Denjoy) *If the coefficients $a_n$, $b_n$ in the trigonometric series (5.1) are all real and if (5.1) converges absolutely for x belonging to a set E of positive measure, then the series (5.49) is convergent.*

*Proof.* We may suppose that $a_0 \geqslant 0$. Write

$$\frac{a_0}{2} = R_0, \quad \sqrt{(a_n^2 + b_n^2)} = R_n, \quad a_n = R_n \cos \alpha_n,$$

$$b_n = R_n \sin \alpha_n \quad (n = 1, 2, \ldots).$$

The the absolute convergence of (5.1) on $E$ implies that

$$\sum_{n=0}^{\infty} R_n |\cos (nx - \alpha_n)| < +\infty, \quad x \in E.$$

Let $A(x) = \sum_{n=0}^{\infty} R_n \cos^2 (nx - \alpha_n)$.

Then, for $x \in E$, $A(x) \leqslant \sum_{n=0}^{\infty} R_n |\cos (nx - \alpha_n)| < +\infty$. Hence $A(x)$ is measurable and finite on the set E. By Egoroff's theorem and a standard argument there is a perfect subset $E_0$ of positive measure such that

$$\sum_{n=0}^{\infty} R_n \int_{E_0} \cos^2 (nx - \alpha_n)\, dx = \int_{E_0} A(x)\, dx < +\infty. \qquad (5.50)$$

Now

$$\int_{E_0} \cos^2 (nx - \alpha_n)\, dx = \frac{1}{2} \int_{E_0} (1 + \cos 2(nx - \alpha_n))\, dx$$

$$= \frac{1}{2} mE_0 + \frac{1}{2} \int_{E_0} \cos 2(nx - \alpha_n)\, dx$$

where $mE_0$ is the measure of $E_0$. Let $\chi$ be the characteristic function of $E_0$, that is,

$$\chi(x) = \begin{cases} 1 & \text{if } x \in E_0 \\ \\ 0 & \text{if } x \notin E_0. \end{cases}$$

Thus,

$$\int_{E_0} \cos 2(nx - \alpha_n)\, dx = \int_{-\pi}^{\pi} \chi(x) \cos 2(nx - \alpha_n)\, dx$$

$$= \cos 2\alpha_n \int_{-\pi}^{\pi} \chi(x) \cos 2nx\, dx +$$

$$+ \sin 2\alpha_n \int_{-\pi}^{\pi} \chi(x) \sin 2nx\, dx$$

$$= o(1)$$

by the Riemann-Lebesgue theorem. Hence, there exists $n_0$ such that for $n > n_0$,

$$\int_{E_0} \cos^2 (nx - \alpha_n)\, dx > \frac{mE_0}{4},$$

and consequently (5.50) implies the convergence of the series $\sum_{n=1}^{\infty} R_n$. Since $|a_n| \leqslant R_n$, $|b_n| \leqslant R_n$ $(n \geqslant 1)$, the theorem is proved. $\triangle$

5.56  THEOREM.  *If the series (5.1) is a Fourier series and*

$$\sum_{k=1}^{n} k \left( |a_k| + |b_k| \right) = o(n) \qquad (5.51)$$

*then the series (5.1) converges almost everywhere (a.e.).  If it is the Fourier series of f, where $f \in C_{2\pi}$, then it converges uniformly over any interval $[A,B]$.*

*Proof.*  $\left| S_n(x) - \sigma_n(x) \right| = \left| \dfrac{1}{n+1} \sum_{k=1}^{n} k (a_k \cos kx + b_k \sin kx) \right|$

$$\leqslant \frac{1}{n+1} \sum_{k=1}^{n} k \left( |a_k| + |b_k| \right)$$

$$= o(1).$$

Hence $S_n(x) - \sigma_n(x) \to 0$. uniformly.  But by Fejér–Lebesgue theorem $\sigma_n(x) \to f(x)$ a.e.   Hence $S_n(x) \to f(x)$ a.e.

If $f \in C_{2\pi}$, then $\sigma_n(x) \to f(x)$ uniformly (see Theorem 5.37) and hence, $S_n(x) \to f(x)$ uniformly on $[A,B]$.   $\triangle$

Note that the condition (5.51) is satisfied if $a_n = o(1/n)$, $b_n = o(1/n)$.

*Exercise.*  Let $f$ be real-valued and in $C_{2\pi}$.        Prove that if $\sum_{1}^{\infty} (1/\sqrt{n})\, \omega\, (1/n)$ is convergent, then the Fourier series (5.10) of $f$ converges absolutely.

*Hint*:  Write (5.1) as $\sum_{k=0}^{\infty} A_k(x)$ and $\rho_n^2 = a_n^2 + b_n^2$.   Then

$$\frac{1}{\pi} \int_{-\pi}^{\pi} \{ f(x+h) - f(x-h) \}^2 \, dx = 4 \sum_{n=1}^{\infty} \rho_n^2 \sin^2 nh.$$

This gives

$$\sum_{n=1}^{\infty} \rho_n^2 \sin^2 \frac{\pi n}{2^{\nu+1}} \leqslant \frac{1}{2} \omega^2 \left( \frac{\pi}{2^{\nu}} \right); \qquad \nu = 1, 2, \ldots.$$

and hence

$$\sum_{n=2^{\nu-1}+1}^{2^{\nu}} \rho_n^2 \leqslant \omega^2 \left( \pi/2^{\nu} \right).$$

Use Schwarz' inequality to get

$$\sum_{n=2}^{\infty} \rho_n \leqslant \sum_{\nu=1}^{\infty} 2^{\nu/2} \, \omega \left( \frac{\pi}{2^{\nu}} \right)$$

and note that our hypothesis implies the convergence of the series on the right.

### Fourier Transforms

In this section we define the Fourier transform of $f$ (where $f \in L (-\infty, \infty)$) and $C^{\alpha}$ summability for integrals and prove the inversion formula of the Fourier transform.

5.57  DEFINITION.  Let $f \in L (-\infty, \infty)$.  The Fourier transform $\hat{f}$ of $f$ is given by

$$\hat{f}(x) = \int_{-\infty}^{\infty} e^{-ixt} f(t) \frac{dt}{\sqrt{(2\pi)}} \qquad (-\infty < x < \infty). \qquad (5.52)$$

Note that the integral on the right exists for all real $x$.
The Fourier transform of $f$ is sometimes defined as

$$\hat{f}(x) = \int_{-\infty}^{\infty} e^{-ixt} f(t) \, dt$$

or as

$$\hat{f}(x) = \int_{-\infty}^{\infty} e^{ixt} f(t) \, dt.$$

The form (5.52) has the advantage of giving the following inversion formula of the Fourier transform in a symmetric form.

If

$$\hat{f}(x) = \int_{-\infty}^{\infty} e^{-ixt} f(t) \frac{dt}{\sqrt{(2\pi)}}$$

then, under suitable conditions,

$$f(t) = \int_{-\infty}^{\infty} e^{itx} \hat{f}(x) \frac{dx}{\sqrt{(2\pi)}} . \qquad (5.53)$$

(See Theorem 5.63 and Corollary 5.65 for a set of conditions under which (5.53) holds.)  Compare the two formulae (5.52) and (5.53) with the formulae for $c_k$ and $f(x)$ given prior to Example 5.5.

5.58  THEOREM.  The Fourier transform $\hat{f}$ is bounded and continuous on $(-\infty, \infty)$.

Proof.  Write $||f||_1 = \int_{-\infty}^{\infty} |f(t)| \frac{dt}{\sqrt{(2\pi)}}$ .  Then, for $-\infty < x < \infty$,

$$|\hat{f}(x)| \leqslant \int_{-\infty}^{\infty} |e^{-ixt} f(t)| \frac{dt}{\sqrt{(2\pi)}}$$

$$= \int_{-\infty}^{\infty} |f(t)| \frac{dt}{\sqrt{(2\pi)}} = ||f||_1 .$$

Also, for any real $x$ and $h$ and for $a > 0$,

$$|\hat{f}(x+h) - \hat{f}(x)| = |\int_{-\infty}^{\infty} e^{-ixt} (e^{-iht}-1) f(t) \frac{dt}{\sqrt{(2\pi)}}|$$

$$\leqslant \int_{-\infty}^{\infty} |f(t)| 2 |\sin \frac{th}{2}| \frac{dt}{\sqrt{(2\pi)}}$$

$$\leqslant 2\int_{-\infty}^{-a} |f(t)| \frac{dt}{\sqrt{(2\pi)}} + 2\int_{a}^{\infty} |f(t)| \frac{dt}{\sqrt{(2\pi)}} +$$

$$+ \int_{-a}^{a} |f(t)| |th| \frac{dt}{\sqrt{(2\pi)}} .$$

Given $\varepsilon > 0$, choose $a$ such that

$$\int_{-\infty}^{-a} |f(t)| dt < \varepsilon \quad \text{and} \quad \int_{+a}^{\infty} |f(t)| dt < \varepsilon.$$

Now, choose $\delta > 0$ such that

$$a\delta \int_{-a}^{a} |f(t)| dt < \varepsilon.$$

Hence, for $|h| < \delta$,

$$|\hat{f}(x+h) - \hat{f}(x)| < \frac{4\varepsilon}{\sqrt{(2\pi)}} + \frac{\delta a}{\sqrt{(2\pi)}} \int_{-a}^{a} |f(t)| dt$$

$$< \frac{5}{\sqrt{(2\pi)}} \varepsilon.$$

Therefore, $\hat{f}$ is continuous at $x$. $\triangle$

5.59 THEOREM. (Riemann–Lebesgue) *If* $f \in L (-\infty,\infty)$, *then*

$$\lim_{x\to\pm\infty} \hat{f}(x) = \lim_{x\to\pm\infty} \int_{-\infty}^{\infty} e^{-ixt} f(t) \frac{dt}{\sqrt{(2\pi)}} = 0.$$

*Proof.* From (5.52) we have

$$-\hat{f}(x) = \int_{-\infty}^{\infty} e^{-ixt} e^{i\pi} f(t) \frac{dt}{\sqrt{(2\pi)}}$$

$$= \int_{-\infty}^{\infty} e^{-ix[t - (\pi/x)]} f(t) \frac{dt}{\sqrt{(2\pi)}}$$

$$= \int_{-\infty}^{\infty} e^{-ixt} f(t +\frac{\pi}{x}) \frac{dt}{\sqrt{(2\pi)}} .$$

Subtracting from (5.52) we have

$$|2\hat{f}(x)| = |\int_{-\infty}^{\infty} e^{-ixt} (f(t) - f(t+\frac{\pi}{x})) \frac{dt}{\sqrt{(2\pi)}}|$$

$$\leqslant \int_{-\infty}^{\infty} |f(t) - f(t+\frac{\pi}{x})| \frac{dt}{\sqrt{(2\pi)}}$$

$$= o(1) \quad (x \to \pm \infty),$$

(see E. Hewitt and K. Stromberg, *Real and Abstract Analysis*, page 199). $\triangle$

5.60 COROLLARY. *If* $f \in L$ $(-\infty,\infty)$ *then*

$$\lim_{x \to \pm\infty} \int_{-\infty}^{\infty} f(t)\sin xt \, dt = \lim_{x \to \pm\infty} \int_{-\infty}^{\infty} f(t)\cos xt \, dt = 0.$$

We now define summability $C^{\alpha}$, where $\alpha > 0$, for integrals and prove the $C^1$ method is regular.

5.61 DEFINITION. *Let* $f$ *be integrable on* $[-T,T]$ *for every* $T > 0$. *The integral* $\int_{-\infty}^{\infty} f(x) \, dx$ *is summable* $C^{\alpha}$ *to the value* $S$ *if*

$$\lim_{T \to \infty} \int_{-T}^{T} (1 - \frac{|x|}{T})^{\alpha} f(x) \, dx = S.$$

5.62 THEOREM. *If* $f(x) \in L$ $(-\infty,\infty)$ *and* $I = \int_{-\infty}^{\infty} f(x) \, dx = S$ *then the integral* $I$ *is summable* $C^1$ *to* $S$.

*Proof.* For each $T > 0$ let

$$f_T(x) = \begin{cases} (1 - \frac{|x|}{T}) f(x), & |x| \leqslant T \\ \\ 0 & |x| > T. \end{cases}$$

Thus,

$$|f_T(x)| \leqslant |f(x)| \quad \text{for} -\infty < x < \infty$$

$$\text{and } \lim_{T \to \infty} f_T(x) = f(x).$$

Hence, by the Lebesgue dominated convergence theorem,

$$\lim_{T \to \infty} \int_{-\infty}^{\infty} f_T(x) \, dx = \int_{-\infty}^{\infty} f(x) \, dx = S,$$

that is,

$$\lim_{T \to \infty} \int_{-T}^{T} \left(1 - \frac{|x|}{T}\right) f(x) \, dx = S. \quad \triangle$$

If an integral $I$ is summable $C^{\alpha}$, where $\alpha > 0$, then it is summable $C^{\beta}$, where $\beta > \alpha$, to the same value (see E. C. Titchmarsh, *Introduction to the theory of Fourier integrals*, pages 27–8).

The next theorem proves that the integral (5.53) is summable $C^1$ to $f(t)$ for almost all $t$.

5.63  THEOREM.  *Let $f \in L \, (-\infty, \infty)$. If $t$ is in the Lebesgue set for $f$ then*

$$\lim_{T \to \infty} \int_{-T}^{T} \left(1 - \frac{|x|}{T}\right) e^{itx} \, \hat{f}(x) \, \frac{dx}{\sqrt{(2\pi)}} = f(t). \tag{5.54}$$

*In particular (5.54) holds for almost all $t \, (-\infty < t < \infty)$.*

*Proof.*  Let $t$ be real, $T > 0$ and denote by $I(t)$ the integral on the left side of (5.54).  Then

$$I(t) = \int_{-T}^{T} \left(1 - \frac{|x|}{T}\right) e^{itx} \left(\int_{-\infty}^{\infty} e^{-iux} \, f(u) \, \frac{du}{\sqrt{(2\pi)}}\right) \frac{dx}{\sqrt{(2\pi)}}.$$

By absolute convergence of the iterated integral, we can change the order of integration and obtain

$$I(t) = \int_{-\infty}^{\infty} f(u) \left(\int_{-T}^{T} \left(1 - \frac{|x|}{T}\right) e^{ix(t-u)} \, \frac{dx}{\sqrt{(2\pi)}}\right) \frac{du}{\sqrt{(2\pi)}}.$$

Now

$$\int_{-T}^{T} \left(1 - \frac{|x|}{T}\right) e^{ix(t-u)} \, \frac{dx}{\sqrt{(2\pi)}} = \int_{0}^{T} \left(1 - \frac{x}{T}\right) 2 \cos x(t-u) \, \frac{dx}{\sqrt{(2\pi)}}$$

$$= \frac{2}{T} \, \frac{1}{(t-u)^2} \, \frac{1 - \cos T(t-u)}{\sqrt{(2\pi)}} \, .$$

Hence

$$I(t) = \frac{1}{\pi} \int_{-\infty}^{\infty} f(u) \, \frac{1 - \cos T(t-u)}{T(t-u)^2} \, du$$

$$= \frac{1}{\pi} \int_{0}^{\infty} \{f(t+y) + f(t-y)\} \, \frac{1 - \cos Ty}{Ty^2} \, dy.$$

Since

$$\int_0^\infty \frac{1-\cos Ty}{Ty^2} \, dy = \int_0^\infty \frac{1-\cos t}{t^2} \, dt = \frac{\pi}{2}$$

we have

$$I(t) - f(t) = \frac{1}{\pi} \int_0^\infty \{ f(t+x) + f(t-x) - 2f(t) \} \frac{1-\cos Tx}{Tx^2} \, dx.$$

The argument now is similar to that in Theorem 5.46. Let

$$\phi(x) = f(t+x) + f(t-x) - 2f(t)$$

and

$$\Phi(x) = \int_0^x |\phi(u)| \, du.$$

Since $t$ is in the Lebesgue set for $f$, given $\varepsilon > 0$, we can find $\eta \ (0 < \eta \leqslant 1)$ such that

$$\Phi(x) < x\varepsilon \quad \text{for } x \in [0,\eta].$$

Fix $\eta$. Since $\int_\eta^\infty \frac{|\phi(x)|}{x^2} \, dx < \infty$, we can find $T_0$ such that for $T > T_0$

$$\frac{2}{T} \int_\eta^\infty \frac{|\phi(x)|}{x^2} \, dx < \varepsilon.$$

Let $T > \max(T_0, 1/\eta)$. Thus

$$|I(t) - f(t)| \leqslant \frac{1}{\pi} \int_0^\infty |\phi(x)| \frac{1-\cos Tx}{Tx^2} \, dx$$

$$= \frac{1}{\pi} \left[ \int_0^{1/T} + \int_{1/T}^\eta + \int_\eta^\infty \right] = \frac{1}{\pi} [I_1 + I_2 + I_3] \quad \text{say.}$$

So

$$I_3 \leqslant \frac{2}{T} \int^\infty \frac{|\phi(x)|}{x^2} \, dx < \varepsilon.$$

Also $1 - \cos Tx \leqslant \frac{1}{2} T^2 x^2$ and so

$$I_2 \leqslant \frac{T}{2} \int_0^{1/T} |\phi(x)| \, dx = \frac{T}{2} \Phi\left(\frac{1}{T}\right) \leqslant \frac{T}{2} \frac{\varepsilon}{T} = \frac{\varepsilon}{2};$$

$$I_2 \leqslant 2 \int_{1/T}^{\eta} \frac{|\phi(x)|\,dx}{Tx^2} = \frac{2}{T}\left[\frac{\Phi(\eta)}{\eta^2} - \Phi\left(\frac{1}{T}\right)T^2 + 2\int_{1/T}^{\eta} \frac{\Phi(x)}{x^3}\,dx\right]$$

$$\leqslant \frac{2}{T}\left(\frac{\eta\varepsilon}{\eta^2} + 2\varepsilon\int_{1/T}^{\eta} \frac{dx}{x^2}\right)$$

$$\leqslant \frac{2}{T}\left(\frac{\varepsilon}{\eta} + 2\varepsilon T\right) < 6\varepsilon.$$

Hence, for $T > \max(T_0, 1/\eta)$,

$$|I(t) - f(t)| \leqslant \frac{1}{\pi}\frac{15\varepsilon}{2}$$

and so

$$\lim_{T \to \infty} I(t) = f(t). \qquad \triangle$$

**5.64  COROLLARY.**  *If $f \in L\ (-\infty,\infty)$ and $f$ is continuous at $t$ then (5.54) holds.*

**5.65  COROLLARY.**  *If $f \in L(-\infty,\infty)$, $\hat{f} \in L(-\infty,\infty)$, and $f$ is continuous at $t$ then*

$$f(t) = \int_{-\infty}^{\infty} e^{ixt}\,\hat{f}(x)\,\frac{dx}{\sqrt{(2\pi)}}.$$

*Proof.*  By Corollary 5.64 the integral on the right is summable $C^1$ to $f(t)$.  But, by hypothesis, the integral is absolutely convergent. Since the $C^1$ method is regular (Theorem 5.62), the corollary follows.$\triangle$

**5.66  COROLLARY.**  (Uniqueness of the Fourier Transform).  *If $f \in L\ (-\infty,\infty)$ $F \in L\ (-\infty,\infty)$, and*

$$\hat{f}(x) = \hat{F}(x) \quad (-\infty < x < \infty)$$

*then*                    $f(t) = F(t)$    *for almost all $t$, $-\infty < t < \infty$.*

*Proof.*  $(f-F)\hat{} = \hat{f} - \hat{F} \equiv 0.$    Hence Theorem 5.63 shows that $f(t) - F(t) = 0$ for almost all $t$.    $\triangle$

*Exercises.*  (i)  Prove that if $f \in L\ (-\infty,\infty)$ and is of bounded variation in some neighborhood of a point $t$ then

$$\lim_{T \to \infty} \int_{-T}^{T} e^{itx}\,\hat{f}(x)\,\frac{dx}{\sqrt{(2\pi)}} = \frac{1}{2}\{f(t+0) + f(t-0)\}.$$

Deduce that if $f$ is continuous at $t$ then

$$f(t) = \lim_{T \to \infty} \int_{-T}^{T} e^{itx} \hat{f}(x) \frac{dx}{\sqrt{(2\pi)}} \, .$$

*Hint*: The proof is similar to that of Jordan's Test for the convergence of a Fourier series [Theorem 5.34].

(ii) If $f \in L(-\infty, \infty)$ and $g \in L(-\infty, \infty)$ then prove that the integral

$$\int_{-\infty}^{\infty} f(x-t)\, g(t) \frac{dt}{\sqrt{(2\pi)}}$$

exists for almost all $x$. Denote it by $h(x)$ whenever the integral exists. $h$ is called the convolution of $f$ and $g$ and is denoted by $f*g$. Prove that $h$ belongs to $L(-\infty, \infty)$ and

$$f*g = g*f,$$

$$\|f*g\|_1 \leqslant \|f\|_1 \|g\|_1,$$

and

$$\hat{h} = \hat{f}\hat{g}.$$

(iii) Verify that

$$f(x) = \sqrt{\left(\frac{2}{\pi}\right)} \frac{1}{1+x^2} \quad \text{and} \quad F(x) = e^{-|x|}$$

are Fourier transforms of each other.

# CHAPTER 6

# Applications of Summability to Analytic Continuation

In the theory of functions of a complex variable we know that a power series, $\sum_{n=0}^{\infty} a_n (z-\alpha)^n$, with positive radius of convergence, $R$, represents an analytic function on the disc $\{z: |z-\alpha| < R\}$, i.e. the sequence of partial sums $\{s_n(z)\}_0^{\infty}$ converges on the disc $\{z: |z-\alpha| < R\}$ to a function $f(z)$ which is analytic on the disc and, conversely, every function $f(z)$ analytic on the disc $\{z: |z-\alpha| < R\}$ $(R > 0)$ can be expanded in the power series $\sum_{n=0}^{\infty} [f^{(n)}(\alpha)/n!] (z-\alpha)^n$ which converges for $|z-\alpha| < R$ and converges uniformly on compact subsets of this domain. Now, we are going to apply a summability transform to the sequence of partial sums of a power series and examine the uniform convergence of this transformation on compact (i.e. closed and bounded) subsets of a domain $D$ containing the original disc of convergence $\{z: |z-\alpha| < R\}$. We will then appeal to

**6.1 THEOREM.** *If $\{g_n(z)\}_0^{\infty}$ is a sequence of functions, each analytic on a domain $D$, and this sequence converges uniformly on compact subsets of $D$ then the limiting function, $g(z)$, is analytic on $D$* (see A. I. Markushevich, *Theory of Functions of a Complex Variable*, Volume 1, page 333).

If our transformation is reasonably well-behaved (for example, if it is regular) then the limiting function $g(z)$ (in Theorem 6.1) will agree with our original function $f(z)$ on the disc $\{z: |z-\alpha| < R\}$ and, hence, $g(z)$ will yield the analytic continuation of $f(z)$ to the domain $D$. In our discussion we let $\alpha = 0$ in order to simplify some of our calculations.

## The Borel Exponential Method

Let us begin by examining the special case of the Borel exponential transformation (see Definition 3.12).

151

6.2 DEFINITION. *Let f be analytic at the origin. If s is a singular point of f then let $\ell(s)$ be the line perpendicular to the line segment $[0,s]$ passing through the point s and let $H_{\ell(s)}$ be the open half-plane containing the origin with boundary $\ell(s)$. The Borel polygon with respect to f is the set*

$$B(f) = \cap H_{\ell(s)}$$

*where the intersection is taken over all singular points of f.*

B(f) *is a domain since it is the intersection of open, arc-wise connected sets.*

6.3 EXAMPLE. The Borel polygon of $f(z) = 1/(1-z)$ is the half-plane $\{z : \operatorname{Re}(z) < 1\}$ since $z = 1$ is the only point of singularity of f.

Before proving the main theorem concerning the Borel exponential method (Theorem 6.5) we need the following preparatory lemma.

6.4 LEMMA. *Let f be analytic at the origin. The function f is analytic inside and on the circle C having diameter $[0,z]$ if and only if $z \in B(f)$.*

*Proof.* First, suppose $z \in B(f)$. If f has a singular point s inside the circle C having diameter $[0,z]$ then we have, by a simple geometric argument, that the perpendicular to $[0,s]$ passing through s intersects $[0,z]$ inside the circle C. Therefore $z \notin B(f)$ which is a contradiction. If f has a singular point s on the circle C then the perpendicular to $[0,s]$ passing through s also passes through z and, hence, $z \notin B(f)$ which, again, is a contradiction.

Now, suppose f is analytic inside and on C. Therefore, there exists a circle $\Gamma$ concentric to C such that $C \subseteq \operatorname{int}(\Gamma)$ and f is analytic on $\operatorname{int}(\Gamma)$. Let s be a singular point of f. There exists a unique point $s_1 \in \Gamma$ such that $\arg s = \arg s_1$. Now, if $\ell(s)$ and $\ell(s_1)$ are the lines perpendicular to $[0,s]$ at s and $[0,s_1]$ at $s_1$ respectively and $H_{\ell(s)}$ and $H_{\ell(s_1)}$ are the open half-planes containing the origin and having, respectively, $\ell(s)$ and $\ell(s_1)$ as boundaries then $z \in H_{\ell(s_1)} \subseteq H_{\ell(s)}$ for any singularity s and, hence, $z \in B(f)$. $\triangle$

6.5 THEOREM. *Let $s_n(z) = \sum\limits_{k=0}^{n} a_k z^k$ where $\sum\limits_{k=0}^{\infty} a_k z^k$ has radius of convergence $R > 0$. If $f(z) = \sum\limits_{k=0}^{\infty} a_k z^k$ on $S_R(0)$ and $z \in B(f)$ then*

$$\lim_{x \to \infty} e^{-x} \sum_{k=0}^{\infty} \frac{x^k}{k!} s_k(z) = f(z).$$

*Proof.* Let $z \in B(f)$. If $z = 0$ the result follows. So assume that $z \neq 0$. By Lemma 6.4 we have that $f$ is analytic inside and on the circle $C$ having diameter $[0,z]$. Let $\Gamma_1$ and $\Gamma$ be circles, each concentric to $C$, where $f$ is analytic on int $(\Gamma_1)$, $C \subseteq$ int $(\Gamma)$, and $\Gamma \subseteq$ int $(\Gamma_1)$. Now,

$$
\begin{aligned}
s_n(z) &= \sum_{k=0}^{n} a_k z^k = \sum_{k=0}^{n} \left( \frac{1}{2\pi i} \int_{\Gamma} \frac{f(t)}{t^{k+1}} \, dt \right) z^k \\
&= \frac{1}{2\pi i} \int_{\Gamma} \frac{f(t)}{t} \left( \sum_{k=0}^{n} \left(\frac{z}{t}\right)^k \right) dt \\
&= \frac{1}{2\pi i} \int_{\Gamma} \frac{f(t)}{t} \left[ \frac{1 - (z/t)^{n+1}}{1 - (z/t)} \right] dt \\
&= \frac{1}{2\pi i} \int_{\Gamma} \frac{f(t)}{t-z} \left( 1 - \left(\frac{z}{t}\right)^{n+1} \right) dt \\
&= \frac{1}{2\pi i} \int_{\Gamma} \frac{f(t)}{t-z} \, dt - \frac{1}{2\pi i} \int_{\Gamma} \frac{f(t)}{t-z} \left(\frac{z}{t}\right)^{n+1} dt \\
&= f(z) - \frac{1}{2\pi i} \int_{\Gamma} \frac{f(t)}{t-z} \left(\frac{z}{t}\right)^{n+1} dt.
\end{aligned}
$$

Therefore,

$$
\lim_{x \to \infty} e^{-x} \sum_{k=0}^{\infty} \frac{x^k}{k!} s_k(z)
$$

$$
= \lim_{x \to \infty} \left\{ e^{-x} \sum_{k=0}^{\infty} \frac{x^k}{k!} f(z) - e^{-x} \sum_{k=0}^{\infty} \frac{x^k}{k!} \left( \frac{1}{2\pi i} \int_{\Gamma} \frac{f(t)}{t-z} \left(\frac{z}{t}\right)^{k+1} dt \right) \right\}
$$

$$
= f(z) - \lim_{x \to \infty} \frac{1}{2\pi i} \int_{\Gamma} \frac{f(t)}{t-z} \left(\frac{z}{t}\right) e^{-x} \left[ \sum_{k=0}^{\infty} \frac{1}{k!} \left(\frac{xz}{t}\right)^k \right] dt
$$

$$
= f(z) - \lim_{x \to \infty} \frac{1}{2\pi i} \int_{\Gamma} \frac{f(t)}{t-z} \left(\frac{z}{t}\right) \exp \left[ x \left(\frac{z}{t} - 1\right) \right] dt.
$$

Now, if Re $\left(\frac{z}{t} - 1\right) \leqslant -\varepsilon < 0$ for some $\varepsilon > 0$ and all $t \in \Gamma$ then

$$
\left| \frac{1}{2\pi i} \int_{\Gamma} \frac{f(t)}{t-z} \left(\frac{z}{t}\right) \exp \left[ x \left(\frac{z}{t} - 1\right) \right] dt \right|
$$

$$
\leqslant \frac{1}{2\pi} \sup_{t \in \Gamma} \left| \frac{z f(t)}{t(t-z)} \right| L(\Gamma) \exp \left\{ \text{Re} \left[ x \left(\frac{z}{t} - 1\right) \right] \right\}
$$

where $L(\Gamma)$ is the length of $\Gamma$. Since there exists $M > 0$ such that

$$
\frac{1}{2\pi} \sup_{t \in \Gamma} \left| \frac{z f(t)}{t(t-z)} \right| L(\Gamma) \leqslant M
$$

we obtain

$$\left| \frac{1}{2\pi i} \int_\Gamma \frac{f(t)}{t-z} \left(\frac{z}{t}\right) \exp\left[x\left(\frac{z}{t} - 1\right)\right] dt \right| \leqslant M \exp(-x\varepsilon)$$

and, hence,

$$\lim_{x\to\infty} \frac{1}{2\pi i} \int_\Gamma \frac{f(t)}{t-z} \left(\frac{z}{t}\right) \exp\left[x\left(\frac{z}{t} - 1\right)\right] dt = 0$$

and the theorem is proved. Therefore, it remains to be shown that there exists $\varepsilon > 0$ such that $\text{Re}[(z/t) - 1] \leqslant - \varepsilon$ for all $t \in \Gamma$.

Let $z = u + iv$ (remember that $z$ is fixed and is non-zero), $t = x + iy$, and $\delta = \inf\{ |\omega - \omega_1| : \omega \in C \text{ and } \omega_1 \in \Gamma \} > 0$. We have $\text{Re}\left(\frac{z}{t}\right) = \frac{ux + vy}{x^2 + y^2}$. The equation of the circle $\Gamma$ is given by

$$\left(x - \frac{u}{2}\right)^2 + \left(y - \frac{v}{2}\right)^2 = \left(\frac{1}{2}\sqrt{u^2 + v^2} + \delta\right)^2 = \frac{1}{4}(u^2 + v^2) + \delta'$$

where $\delta' = \delta \sqrt{u^2 + v^2} + \delta^2$. Therefore

$$\frac{ux + vy}{x^2 + y^2} = 1 - \frac{\delta'}{x^2 + y^2},$$

i.e.
$$\text{Re}\left(\frac{z}{t}\right) = 1 - \frac{\delta'}{x^2 + y^2}.$$

Choose $\varepsilon = \delta'/[(|z| + \delta)^2]$. Therefore

$$0 < \varepsilon < 1 \quad \text{and} \quad 1 - \frac{\delta'}{x^2 + y^2} \leqslant 1 - \frac{\delta'}{(|z| + \delta)^2}.$$

So $\text{Re}(z/t) \leqslant 1 - \varepsilon$, i.e. $\text{Re}[(z/t) - 1] \leqslant - \varepsilon < 0$. $\triangle$

We see that the Borel exponential transform of the sequence of partial sums of $\sum_{k=0}^\infty a_k z^k$ gives us the analytic continuation of the power series $\sum_{k=0}^\infty a_k z^k$ to the Borel polygon $B(f)$ (since $B(f)$ is a domain and $S_R(0) \subset B(f)$).

*Exercise.* Paraphrase Theorem 6.5 using the Borel matrix transformation $[b_{n,k} = e^{-n} (n^k/k!)]$ instead and prove your theorem.

## The Okada Theorem

Now, we want to examine a more general setting, in particular, where we will use a general matrix transformation of the sequence of partial sums of $\sum_{k=0}^{\infty} a_k z^k$ (which has positive radius of convergence).

In the following material let $A = (a_{n,k})$ be the matrix under consideration and let $\{s_k(z)\}_0^{\infty}$ be the sequence of partial sums of the geometric series, i.e.

$$s_k(z) = \sum_{n=0}^{k} z_k = \frac{1 - z^{k+1}}{1 - z}, \quad z \neq 1.$$

Furthermore, let $R^1 = \{z: \lim_{n \to \infty} \sum_{k=0}^{\infty} a_{n,k} \, s_k(z) = 1/(1-z)\}$ and assume that

(i) $\{z: |z| < 1\} \subseteq R^1$,

(ii) $R^1$ is open, and

(iii) $\lim_{n \to \infty} \sum_{k=0}^{\infty} a_{n,k} \, s_k(z) = 1/(1-z)$ uniformly on compact subsets of $R^1$.

We, of course, have $1 \notin R^1$.

6.6 LEMMA. *For each fixed $n$ $(n = 0, 1, \ldots)$ the sum $\sum_{k=0}^{\infty} a_{n,k} \, s_k(z)$ converges uniformly on compact subsets of $R^1$.*

*Proof.* Since

$$\sum_{k=0}^{\infty} a_{n,k} \, s_k(z) = \frac{1}{1-z} \sum_{k=0}^{\infty} a_{n,k} - \frac{1}{1-z} \sum_{k=0}^{\infty} a_{n,k} z^{k+1},$$

$(1-z)^{-1}$ is bounded on compact subsets of $R^1$, and $\sum_{k=0}^{\infty} a_{n,k} z^{k+1}$ converges uniformly on compact subsets of $R^1$ we have that $\sum_{k=0}^{\infty} a_{n,k} s_k(z)$ converges uniformly on compact subsets of $R^1$.   △

We will now let $r_n(z) = \sum_{k=0}^{\infty} a_{n,k} \, s_k(z)$ (for $n = 0, 1, \ldots$).

6.7 LEMMA. (i) $\lim_{n \to \infty} r_n(0) = 1$ *and*

(ii)        $\lim_{n \to \infty} r_n^{(k)}(0)/k! = 1$   *for each* $k = 1, 2, \ldots$.

*Proof.* (i) From the assumption that $\lim_{n \to \infty} r_n(z) = 1/(1-z)$ on $R^1$ we have that $\lim_{n \to \infty} r_n(0) = 1$.

(ii) Since $\lim_{n \to \infty} r_n(z) = 1/(1+z)$ uniformly on compact subsets of $R^1$ we have that

$$\frac{d^k}{dz^k}\left(\frac{1}{1-z}\right) = \lim_{n\to\infty} \tau_n^{(k)}(z)$$

uniformly on compact subsets of $R^1$, i.e.

$$\lim_{n\to\infty} \tau_n^{(k)}(z) = \frac{k!}{(1-z)^{k+1}}.$$

Therefore, $\lim_{n\to\infty} \tau_n^{(k)}(0) = k!$, i.e. $\lim_{n\to\infty} \tau_n^{(k)}(0)/k! = 1.$  $\triangle$

Since $\sum_{k=0}^{\infty} a_{n,k} = \tau_n(0)$ we have that condition (ii) of the Silverman–Toeplitz theorem (Theorem 2.3) is satisfied. Also,

$$a_{n,k} = \frac{\tau_n^{(k)}(0)}{k!} - \frac{\tau_n^{(k+1)}(0)}{(k+1)!}$$

and, hence, condition (i) of Theorem 2.3 is satisfied.  Condition (iii) of the Silverman–Toeplitz theorem need not hold, however, to complete our analysis of the problem.  Therefore, we need not restrict our attention to regular matrices.

Now, let $R = \{\omega = (1/z): z \in C(R^1)\} \cup \{0\}$ ($C(R^1)$ is the complement of $R^1$).  Thus, $R \subseteq \{z: |z| \leqslant 1\}$, $[0,1] \subseteq R$ and $R$ is closed.

**6.8  EXAMPLE.**    For    the    Borel    matrix    method    we    have $R = \{z: |z - \frac{1}{2}| \leqslant \frac{1}{2}\}$ since

$$\lim_{n\to\infty} \sum_{k=0}^{\infty} e^{-n}\frac{n^k}{k!} s_k(z) = \lim_{n\to\infty}\left(\frac{1}{1-z} - e^{-n}\frac{z}{1-z}\sum_{k=0}^{\infty}\frac{(nz)^k}{k!}\right) = \frac{1}{1-z}$$

if and only if $\text{Re}(z) < 1$.  Therefore $R^1 = \{z: \text{Re}(z) < 1\}$ and, hence $C(R^1) = \{z: \text{Re}(z) \geqslant 1\}$ and, under the mapping $\omega = 1/z$, we obtain

$$R = \{z: |z - \tfrac{1}{2}| \leqslant \tfrac{1}{2}\}.$$

**6.9  LEMMA.**  *Let $B$ be a set of complex numbers.  If $0 \in B$ ($B \neq \emptyset$) then $\bigcap_{c \notin B} cR^1 = \{c: cR \subseteq B\}$ where $cR = \{cb: b \in R\}$.*

*Proof.*    First, since $0 \in B$, we have that $0R = \{0\} \subseteq B$ and, hence, $0 \in \{c: cR \subseteq B\}$.  Also, $0 \in \{z: |z| < 1\} \subseteq R^1$ so $0 \in cR^1$ for all $c \notin B$.  Therefore $0 \in \bigcap_{c \notin B} cR^1$.

Now, assume that $z \neq 0$.  We have that $z \in \bigcap_{c \notin B} cR^1$ if and only if $z \in cR^1$ for every $c \notin B$, that is, $z/c \in R^1$ for every $c \notin B$ (note that $c \neq 0$ since $0 \in B$).  Now, $z/c \in R^1$ for every $c \notin B$ if and only if

$z/c \notin C(R^1)$ for every $c \notin B$, i.e. $c/z \notin R$ for every $c \notin B$. This last statement is equivalent to $c \notin zR$ for every $c \notin B$, i.e. $zR \subseteq B$ or $z \in \{c: cR \subseteq B\}$.  $\triangle$

We now introduce the notation to be used for the remainder of this chapter. Let $D$ be a fixed, simply connected domain (where $0 \in D$ and $D \neq \mathcal{C}$) and let $\mathcal{F} = \{f: f$ is analytic on $D\}$. Denote the power series representation (about the origin) for $g \in \mathcal{F}$ by

$$\sum_{k=0}^{\infty} b_k(g)\, z^k.$$

Let

$$U(D) = \{z: \lim_{n \to \infty} \sum_{k=0}^{\infty} a_{n,k} \left( \sum_{j=0}^{k} b_j(g) z^j \right) = g(z) \text{ for all } g \in \mathcal{F} \}.$$

We now state and prove two lemmas which we will summarize (along with our notation) in the statement of the Okada theorem (Theorem 6.13).

6.10  LEMMA.  $U(D) = \bigcap_{c \notin D} cR^1 = \{c: cR \subseteq D\}.$

Proof.  By Lemma 6.9 we have the second equality. Next we prove that $\{c: cR \subseteq D\} \subseteq U(D)$. Let $z_0 \in \{c: cR \subseteq D\}$. Thus $z_0 R \subseteq D$. $R$ is compact (since $R \subseteq \{z: |z| \leqslant 1\}$ and $C(R^1)$ is closed) thus $z_0 R$ is compact. Since $D$ is a domain (hence it is open) there exists a simple closed rectifiable curve $\Gamma$ such that $z_0 R \subseteq \text{int}(\Gamma)$ and $\Gamma \subseteq D$. Also, $1 \notin R^1$ so $1 \in R$. Thus $z_0 \in z_0 R \subseteq \text{int}(\Gamma)$. Therefore, if $f \in \mathcal{F}$ we have that

$$f(z_0) = \frac{1}{2\pi i} \int_\Gamma \frac{f(t)}{t-z_0}\, dt = \frac{1}{2\pi i} \int_\Gamma \frac{f(t)}{t} \frac{1}{1 - (z_0/t)}\, dt.$$

Now $t \in \Gamma$ so $t \notin z_0 R$ which implies that $t/z_0 \notin R$, i.e. $\dfrac{t}{z_0} \in C(R)$ and hence, $z_0/t \in R^1$.  Therefore

$$\lim_{n \to \infty} \tau_n \left( \frac{z_0}{t} \right) = \frac{1}{1 - (z_0/t)}$$

uniformly on $\Gamma$ since $\Gamma$ is a compact subset of $R^1$.

So

$$f(z_0) = \frac{1}{2\pi i} \int_\Gamma \frac{f(t)}{t} \left(\lim_{n\to\infty} \tau_n \left(\frac{z_0}{t}\right)\right) dt$$

$$= \lim_{n\to\infty} \frac{1}{2\pi i} \int_\Gamma \frac{f(t)}{t} \tau_n \left(\frac{z_0}{t}\right) dt$$

$$= \lim_{n\to\infty} \frac{1}{2\pi i} \int_\Gamma \frac{f(t)}{t} \sum_{k=0}^\infty a_{n,k} s_k \left(\frac{z_0}{t}\right) dt$$

$$= \lim_{n\to\infty} \sum_{k=0}^\infty a_{n,k} \left(\frac{1}{2\pi i} \int_\Gamma \frac{f(t)}{t} s_k \left(\frac{z_0}{t}\right) dt\right)$$

by Lemma 6.6.  Continuing we obtain

$$f(z_0) = \lim_{n\to\infty} \sum_{k=0}^\infty a_{n,k} \left(\frac{1}{2\pi i} \int_\Gamma \frac{f(t)}{t} \sum_{j=0}^k \left(\frac{z_0}{t}\right)^j dt\right)$$

$$= \lim_{n\to\infty} \sum_{k=0}^\infty a_{n,k} \left(\sum_{j=0}^k \left[\frac{1}{2\pi i} \int_\Gamma \frac{f(t)}{t^{j+1}} dt\right] z_0^j\right)$$

$$= \lim_{n\to\infty} \sum_{k=0}^\infty a_{n,k} \left(\sum_{j=0}^k b_j(f) z_0^j\right)$$

and, thus, $z_0 \in U(D)$.

Finally, we show that $U(D) \subseteq \cap_{c \notin D} cR^1$.  Suppose that $z_0 \notin \cap_{c \notin D} cR^1$.  Thus $z_0 \notin c_0 R^1$ for some $c_0 \notin D$, i.e. $z_0/c_0 \notin R^1$ for some $c_0 \notin D$.  Therefore the sequence of partial sums of $\sum_{k=0}^\infty (z_0/c_0)^k$ is not $A$-summable to

$$\frac{1}{1-(z_0/c_0)} = \frac{c_0}{c_0-z_0}.$$

Let $\phi(z) = c_0/(c_0-z)$.  Since $c_0 \notin D$ we have that $\phi$ is analytic on $D$ and, hence, by the definition of $U(D)$ we have $z_0 \notin U(D)$.  $\triangle$

Before we continue we remark that if $R^1$ is star-like with respect to the origin (i.e. if $z \in R^1$ then $[0,z] \subseteq R^1$) then $U(D)$ is star-like with respect to the origin.  The proof of this statement follows from the fact that $U(D) = \cap_{c \notin D} cR^1$ (by Lemma 6.10) and if $z \in cR^1$ then $z/c \in R^1$ (which implies that $[0,z/c] \subseteq R^1$, i.e. $[0,z] \subseteq cR^1$.  So $U(D)$ is the intersection of sets each star-like with respect to the origin.

We now need a topological result prior to the statement and proof of Lemma 6.12.  It is a well-known result and can be found in any book on introductory topology (e.g. see J. L. Kelley, *General Topology*).

6.11   LEMMA.   *If G is compact and $\mathcal{C}$ is a collection of open sets where $G \subseteq \cup_{A \in \mathcal{C}} A$ then there exists a finite collection $\mathcal{C}^1 \subseteq \mathcal{C}$ such that $G \subseteq \cup_{A \in \mathcal{C}^1} A$.*
We now state the important

6.12   LEMMA.   *If $f \in \mathcal{F}$ then the A-transform of the sequence of partial sums of $\sum\limits_{k=0}^{\infty} b_k(f) z^k$ is uniformly convergent on compact subsets of $U(D)$.*

*Proof.* Let $F$ be a compact subset of $U(D)$ and $f \in \mathcal{F}$. First, we observe that $U(D) \subseteq D$ since, by Lemma 6.10, $U(D) = \{c \colon cR \subseteq D\}$ and $1 \in R$ thus if $z_0 \in U(D)$ then $z_0 \in z_0 R \subseteq D$. Now, for $z_0 \in F$ choose $\delta_0 = \delta(z_0) > 0$ such that $\{z \colon |z-z_0| < \delta_0\} \subseteq D$. Let $S(\delta_0)(z_0) = \{z \colon |z-z_0| < \delta_0\}$. With each $S(\delta_0)(z_0)$ correspond to it the closed neighbourhood (of $z_0$) $N(\delta_0)(z_0) = \{z \colon |z-z_0| \leqslant \delta_0/M_0\}$ where $M_0 > 1$ is taken large enough so that if $\omega \in N(\delta_0)(z_0)$ then $\omega R \subseteq D$. Now $\{\text{int } (N(\delta_0)(z_0)) \colon z_0 \in F\}$ is a collection of open sets where $F \subseteq \cup_{z_0 \in F} [\text{int } (N(\delta_0)(z_0))]$. Therefore, by Lemma 6.11, there exists a finite subcollection (say $\{\text{int } (N(\delta_1)(z_1)), \ldots, \text{int } (N(\delta_M)(z_M))\}$) such that $F \subseteq \cup\limits_{j=1}^{M} [\text{int } (N(\delta_j)(z_j))] \subseteq \cup\limits_{j=1}^{M} N(\delta_j)(z_j)$.

If we can show that

$$\sigma_n{}^f(\omega) = \sum_{k=0}^{\infty} a_{n,k} \sum_{j=0}^{\infty} b_j(f)\omega^j$$

converges uniformly on any $N(\delta_0)(z_0)$ then we are finished (since $F$ is a subset of the union of a finite number of these sets).

By our choice of $\delta_0$ we have that $f$ is analytic on $N(\delta_0)(z_0)$ (which is closed) since $N(\delta_0)(z_0) \subseteq S(\delta_0)(z_0) \subseteq D$. So, there exists a simple closed rectifiable curve $\Gamma$ such that $N(\delta_0)(z_0) \subseteq \text{int }(\Gamma)$ and $\Gamma \subseteq D$ (in fact, we may assume that if $\omega \in N(\delta_0)(z_0)$ then $\omega R \subseteq \text{int}(\Gamma)$). For any $\omega \in N(\delta_0)(z_0)$ we have

$$\sigma_n{}^f(\omega) = \sum_{k=0}^{\infty} a_{n,k} \sum_{j=0}^{k} b_j(f)\omega^j$$

$$= \sum_{k=0}^{\infty} a_{n,k} \left( \sum_{j=0}^{k} \left( \frac{1}{2\pi i} \int_{\Gamma} \frac{f(t)}{t^{j+1}} dt \right) \omega^j \right)$$

$$= \sum_{k=0}^{\infty} a_{n,k} \left[ \frac{1}{2\pi i} \int_{\Gamma} \frac{f(t)}{t} \sum_{j=0}^{k} \left( \frac{\omega}{t} \right)^j dt \right]$$

$$= \sum_{k=0}^{\infty} a_{n,k} \left[ \frac{1}{2\pi i} \int_{\Gamma} \frac{f(t)}{t} s_k \left( \frac{\omega}{t} \right) dt \right]$$

$$= \frac{1}{2\pi i} \int_{\Gamma} \frac{f(t)}{t} \left[ \sum_{k=0}^{\infty} a_{n,k} s_k \left( \frac{\omega}{t} \right) \right] dt.$$

Note that $\{ \frac{\omega}{t} : \omega \in N(\delta_0)(z_0)$ and $t \in \Gamma \}$ is a subset of $R^1$ since $\omega/t \in R^1 (\omega R \subseteq \text{int}(\Gamma)$ thus $t \notin \omega R$ which implies that $t/\omega \notin R)$. Thus, $\{ \omega/t : \omega \in N(\delta_0)(z_0)$ and $t \in \Gamma \}$ is a compact subset of $R^1$ and, hence, we have that $\sum_{k=0}^{\infty} a_{n,k} s_k (\omega/t)$ converges uniformly on this set (by Lemma 6.6). Therefore the interchange of the order of integration and summation is permissible.

Therefore

$$\sigma_n^{f}(\omega) = \frac{1}{2\pi i} \int_{\Gamma} \frac{f(t)}{t} \tau_n \left( \frac{\omega}{t} \right) dt$$

and

$$\lim_{n \to \infty} \sigma_n^{f}(\omega) = \lim_{n \to \infty} \frac{1}{2\pi i} \int_{\Gamma} \frac{f(t)}{t} \tau_n \left( \frac{\omega}{t} \right) dt$$

$$= \frac{1}{2\pi i} \int_{\Gamma} \frac{f(t)}{t} \left[ \lim_{n \to \infty} \tau_n \left( \frac{\omega}{t} \right) \right] dt$$

which is uniform on compact subsets of $R^1$. Therefore $\{ \sigma_n^{f}(\omega) \}_0^{\infty}$ converges uniformly on $N(\delta_0)(z_0)$. $\triangle$

We see that, by Theorem 6.1, the $A$-transform of the sequence of partial sums of $\sum_{k=0}^{\infty} b_k(f) z^k$ gives us the analytic continuation to $U(D)$.

We conclude this chapter by summarizing our results in the statement of

6.13 THEOREM. (Okada Theorem) *Let* $A = (a_{n,k})$ *be an infinite matrix of complex entries and denote the kth partial sum of the geometric series,* $\sum_{j=0}^{\infty} z^j$, *by* $s_k(z)$. *Let*

$$R^1 = \{ z : \lim_{n \to \infty} \tau_n(z) = \lim_{n \to \infty} \sum_{k=0}^{\infty} a_{n,k} s_k(z) = \frac{1}{1-z} \}$$

*and assume that* $\{ z : |z| < 1 \} \subseteq R^1$, $R^1$ *is open, and* $\lim_{n \to \infty} \tau_n(z) = 1/(1-z)$ *uniformly on compact subsets of* $R^1$. *Let* $R = \{ \omega = 1/z : z \in C(R^1) \} \cup \{0\}$. *Let D be a fixed domain where* $0 \in D$ *and denote by* $\mathcal{F}$ *the collection of all functions analytic on D. For*

$f \in \mathcal{F}$ denote the power series representation of $f$ (about the origin) by $\sum_{k=0}^{\infty} b_k(f) z^k$.

If $U(D) = \{z: \lim_{n\to\infty} \sigma_n{}^f(z) = \lim_{n\to\infty} \sum_{k=0}^{\infty} a_{n,k} \left( \sum_{j=0}^{k} b_j(f)z^j \right) = f(z)$

for all $f \in \mathcal{F}\}$

then  (i)  $U(D) = \cap_{c \notin D} cR^1 = \{c: cR \subseteq D\}$

and  (ii)  $\lim_{n\to\infty} \sigma_n{}^f(z)$ converges uniformly on compact subsets of $U(D)$.

This result is essentially a theorem due to Y. Okada in the article 'Uber die Annaherung analytischer Funktionen', *Math. Z.* volume 23, pages 62–71 published in 1925.

To re-emphasize the importance of the Okada theorem assume that we are given a power series $\sum_{k=0}^{\infty} b_k z^k$ with positive radius of convergence a summability transform $A = (a_{n,k})$, and that we know the existence of a domain $D$ $(0 \in D)$ into which the power series can be analytically continued.  Suppose we have the following information concerning the $A$-transform of the sequence of partial sums of the geometric series $\{\tau_n(z)\}_0^{\infty}$ $\left[ \text{where } \tau_n(z) = \sum_{k=0}^{\infty} a_{n,k} (1-z^{k+1})/(1-z) \right]$ :

$$R^1 = \{z: \lim_{n\to\infty} \tau_n(z) = \frac{1}{1-z}\},$$

$$\{z: |z| < 1\} \subseteq R^1,$$

$$R^1 \text{ is open,}$$

and $\lim_{n\to\infty} \tau_n(z) = 1/(1-z)$ uniformly on compact subsets of $R^1$.  If $\{\sigma_n(z)\}_0^{\infty}$ is the $A$-transform of the sequence of partial sums of the series $\sum_{k=0}^{\infty} b_k z^k$, i.e.

$$\sigma_n(z) = \sum_{k=0}^{\infty} a_{n,k} \sum_{j=0}^{k} b_j z^j,$$

then $\{\sigma_n(z)\}_0^{\infty}$ gives us the analytic continuation of the power series $\sum_{k=0}^{\infty} b_k z^k$ into a domain which contains, as a subset, $U(D)$, i.e. $U(D)$ is a 'minimal' domain into which $\sum_{k=0}^{\infty} b_k z^k$ is continued analytically by means of the summability transform $A = (a_{n,k})$.

*Exercise.* Let $A = (a_{n,k})$ be the Borel matrix method. By Example 6.8 we know that $R^1 = \{z: \text{Re}(z) < 1\}$. Suppose we know that $\sum_{k=0}^{\infty} b_k z^k$ can be analytically continued into the domain $\{z: |z-1| < 2\}$. Letting $D = \{z: |z-1| < 2\}$ show that $U(D)$ is the interior of the ellipse $[(u-1)^2/4] + (v^2/3) = 1$ (where $\omega = u + iv$).

# Appendix

To give an alternative proof to the last part of the proof of Theorem 2.3 (see Theorem A.5) by the use of functional analytic methods, we first define what is meant by a norm, a Banach space, and a continuous linear mapping on normed linear spaces. We then state two theorems, without proofs, which will enable us to prove Theorem A.5. For a complete discussion of these ideas and for the proofs of Theorems A.3 and A.4 we refer the reader to the text *Linear Operators. Part I: General Theory* by N. Dunford and J. Schwartz.

A.1  DEFINITION. *Let $B$ be a vector space (linear space) over the complex numbers $\mathcal{C}$. Let $R^+$ denote the non-negative real numbers.*

(i)  *A norm $N$ on the space $B$ is a function from $B$ into $R^+$ such that*

(a)  $N(x+y) \leqslant N(x) + N(y)$ *for all* $x, y \in B$,

(b)  $N(\alpha x) = |\alpha| N(x)$ *for all* $x \in B$, $\alpha \in \mathcal{C}$, *and*

(c)  $N(x) = 0$ *if and only if* $x = 0$.

(ii)  *The space $B$ is complete in the norm $N$ if every Cauchy sequence in $B$ (with respect to $N$) has its limit in $B$.*

(iii)  *The space $B$ with the norm $N$ (written $(B,N)$) is a Banach space if it is a complete normed linear space.*

It is readily shown that the space $\mathcal{C}$ of complex numbers with norm $|\ |$ (absolute value) is a Banach space, and that $\Omega$ (the collection of all convergent complex sequences) with the norm $||z|| = \sup_n \{|z_n|\}$, where $z = \{z_n\}_0^\infty$, is a Banach space.

A.2  DEFINITION. *Let $(B_1, N_1)$ and $(B_2, N_2)$ be normed linear spaces. A mapping $T$ from $B_1$ to $B_2$*

(i)  *is linear if $T(\alpha x + y) = \alpha T(x) + T(y)$ for all $x, y \in B_1$ and $\alpha \in \mathcal{C}$.*

(ii)  *is continuous at $x_0 \in B_1$ if given $\varepsilon > 0$ there exists $\delta(\varepsilon) > 0$ such that $N_2(T(x) - T(x_0)) < \varepsilon$ whenever $N_1(x - x_0) < \delta(\varepsilon)$.*

**A.3 THEOREM.** *Let T be a linear mapping from $(B_1, N_1)$ to $(B_2, N_2)$ (each of the two spaces are normed linear spaces). Then T is continuous on $B_1$ if and only if there exists $M > 0$ such that $N_2(T(x)) \leqslant MN_1(x)$ whenever $x \in B_1$.*

**A.4 THEOREM.** (Uniform Boundedness Principle) *Let $(B_1, N_1)$ and $(B_2, N_2)$ be Banach spaces. Let $\{T_n\}_0^\infty$ be a sequence of linear continuous functions, each from $B_1$ to $B_2$, such that $\{T_n(x)\}_0^\infty$ is bounded in $B_2$ whenever $x \in B_1$. If $\overline{N}(T_n)$ is defined to be $\inf\{M: x \in B_1 \text{ and } N_2(T_n(x)) \leqslant MN_1(x)\}$ then $\{\overline{N}(T_n)\}_0^\infty$ is a bounded sequence of non-negative numbers.*

We are now able to state and prove (by use of the Uniform Boundedness Principle)

**A.5 THEOREM.** *If $A = (a_{n,k})$ is regular and $\sum\limits_{k=0}^\infty |a_{n,k}|$ converges for each $n = 0, 1, \ldots$ then $\sup_n \{\sum\limits_{k=0}^\infty |a_{n,k}|\} \leqslant M < \infty$ for some $M > 0$.*

*Proof.* Let $\Omega$ be the collection of convergent complex sequences $\{z_k\}_0^\infty$ with norm $\|\ \|$ (if $z = \{z_k\}_0^\infty$ then $\|z\| = \sup_n \{|z_n|\}$). Define the linear mapping $T_n$ from $\Omega$ to $\mathcal{C}$ by

$$T_n(z) = \sum_{k=0}^\infty a_{n,k} z_k \text{ (where } z = \{z_k\}_0^\infty) \quad \text{for each } n = 0, 1, \ldots.$$

To show that $T_n$ is continuous on $\Omega$ let $z \in \Omega$. We have

$$|T_n(z)| = |\sum_{k=0}^\infty a_{n,k} z_k| \leqslant \sum_{k=0}^\infty |a_{n,k}|\,|z_k| \leqslant \left(\sum_{k=0}^\infty |a_{n,k}|\right)\|z\|.$$

Since $\sum\limits_{k=0}^\infty |a_{n,k}|$ converges we have, by Theorem A.3, that $T_n$ is continuous for each $n = 0, 1, \ldots$ . Also, we have that $\overline{N}(T_n) \leqslant \sum\limits_{k=0}^\infty |a_{n,k}|$. We want to prove that $\overline{N}(T_n) = \sum\limits_{k=0}^\infty |a_{n,k}|$. Let $\varepsilon > 0$ be given. Choose a positive integer $K = K(\varepsilon)$ such that $\sum\limits_{k=0}^\infty |a_{n,k}| - \varepsilon < \sum\limits_{k=0}^K |a_{n,k}|$. Define the sequence $y = \{y_j\}_0^\infty$ by $y_j = |a_{n,j}|/a_{n,j}$ if $0 \leqslant j \leqslant K$ and $a_{n,j} \neq 0$, and $y_j = 0$ otherwise. So $y \in \Omega$ (it converges to 0) and $\|y\| \leqslant 1$.

Therefore,

$$\sum_{k=0}^{\infty} |a_{n,k}| - \varepsilon < \sum_{k=0}^{K} |a_{n,k}| = \sum_{k=0}^{K} a_{n,k} y_k = \sum_{k=0}^{\infty} a_{n,k} y_k = T_n(y)$$

$$\leqslant |T_n(y)| \leqslant \overline{N}(T_n) \, \|y\| \leqslant \overline{N}(T_n),$$

i.e.

$$\sum_{k=0}^{\infty} |a_{n,k}| - \varepsilon \leqslant \overline{N}(T_n)$$

and, hence, $\overline{N}(T_n) = \sum_{k=0}^{\infty} |a_{n,k}|.$

To complete the proof let $z \in \Omega$. Since $A = (a_{n,k})$ is regular we have that the sequence $\{T_n(z)\}_0^{\infty} = \{ \sum_{k=0}^{\infty} a_{n,k} z_k \}_0^{\infty}$ converges, in particular, the sequence is bounded. By Theorem A.4, the sequence $\{\overline{N}(T_n)\}_0^{\infty}$ is bounded, i.e. the sequence $\{ \sum_{k=0}^{\infty} |a_{n,k}| \}_0^{\infty}$ is bounded. $\triangle$

# Bibliography

*Following is a list of books which the interested reader will find useful. Included in this list are those books which were referred to within the text.*

1. ALEXITS, G. *Convergence problems of orthogonal series,* Pergamon, New York (1961)

2. ASPLUND, E. and BUNGART, L. *A first course in integration,* Holt, Rinehart and Winston, New York (1966)

3. BARY, N. K. *A treatise on trigonometric series,* Volumes 1 and 2, English translation by M. F. Mullins, Pergamon, New York (1964)

4. BOAS, R. P. *Integrability theorems for trigonometric transforms,* Springer-Verlag, New York (1967)

5. BOCHNER, S. and CHANDRASEKHARAN, K. *Fourier transforms,* Princeton University Press, Princeton (1949)

6. CHANDRASEKHARAN, K. and MINAKSHISUNDARAM, S. *Typical Means,* Tata Institute Monograph, Oxford University Press, London (1952)

7. COOKE, R. G. *Infinite matrices and sequence spaces,* Macmillan, New York (1950)

8. DUNFORD, N. and SCHWARTZ, J. *Linear operators.* Part I: *General theory,* Wiley, New York (1958)

9. EDWARDS, R. E. *Fourier series: a modern introduction,* Volumes 1 and 2, Holt, Rinehart and Winston, New York (1967)

10. GOLDBERG, R. R. *Fourier transforms,* Cambridge University Press, Cambridge (1961)

11. HARDY, G. H. *Divergent series,* Oxford University Press, London (1956)

12. HARDY, G. H. and RIESZ, M. *The general theory of Dirichlet's series,* Cambridge University Press, Cambridge (1915)

13. HARDY, G. H. and ROGOSINSKI, W. W. *Fourier series*, Cambridge University Press, Cambridge (1962)

14. HEWITT, E. and STROMBERG, K. *Real and abstract analysis*, Springer–Verlag, New York (1965)

15. JEFFERY, R. L. *Trigonometric series*, University of Toronto Press, Toronto (1956)

16. KAHANE, J. P. and SALEM, R. *Ensembles parfaits et series trigonometriques*, Herman, Paris (1963)

17. KATZNELSON, Y. *An introduction to harmonic analysis*, Wiley, New York (1968)

18. KELLEY, J. L. *General Topology*, Van Nostrand, New York (1955)

19. KNOPP, K. *Theory and application of infinite series*, English translation by R. C. Young, Blackie, London (1928)

20. MARKUSHEVICH, A. I. *Theory of functions of a complex variable*, Volumes 1–3, English translation by R. A. Silverman, Prentice-Hall, New Jersey (1965)

21. MEINARDUS, G. *Approximation of functions: theory and numerical methods*, English translation by L. L. Schumaker, Springer-Verlag, New York (1967)

22. NATANSON, I. P. *Theory of functions of a real variable*, Volumes 1 and 2, English translation by L. F. Boron, Frederick Ungar, New York (1955)

23. NATANSON, I. P. *Constructive function theory*, Volumes 1–3, English translation, Frederick Ungar, New York (1961)

24. PETERSEN, G. M. *Regular matrix transformations*, McGraw-Hill, London (1966)

25. PEYERIMHOFF, A. *Lectures on summability*, Springer-Verlag, New York (1969)

26. PITT, H. R. *Tauberian theorems*, Oxford University Press, London (1958)

27. ROGOSINSKI, W. *Fourier series*, Chelsea, New York (1950)

28. RUDIN, W. *Real and complex analysis*, McGraw-Hill, New York (1966)

29. SZASZ, O. *Introduction to the theory of divergent series*, Stechert-Hafner, New York (1952)

30. TITCHMARSH, E. C. *The theory of functions*, Oxford University Press, London (1944)

31. TITCHMARSH, E. C. *An introduction to the theory of Fourier integrals*, Oxford University Press, London (1948)

32. TOLSTOV, G. P. *Fourier series*, Prentice-Hall, Englewood Cliffs, New Jersey (1962)

33. WIENER, N. *The Fourier integral and certain of its applications*, Cambridge University Press, Cambridge (1933)

34. ZELLER, K. *Theorie der Limitierungsverfahren*, Springer-Verlag, Berlin (1958)

35. ZYGMUND, A. *Trigonometric series*, Volumes 1 and 2, Cambridge University Press, Cambridge (1959)

**References within the book**

# LIST OF SYMBOLS

| Symbol | | Page where symbol first appears |
|---|---|---|
| $L[-\pi,\pi)$ | space of integrable functions, in the Lebesgue sense, on $[-\pi,\pi)$ | 97 |
| $L^2[a,b)$ | space of functions measurable on $[a,b)$ with integrable square | 102 |
| $L^p[a,b)$ | space of functions measurable on $[a,b)$ with $p$th power integrable | 111 |
| $L_w^2[a,b)$ | space of functions, $f$, measurable on $[a,b)$ with $w|f|^2$ integrable | 106 |
| $L(\Gamma)$ | length of the curve $\Gamma$ | 153 |
| $\mathrm{Lip}_c(\alpha)$ | collection of functions satisfying the Lipschitz condition of order $\alpha$ on $[-\pi,\pi)$ | 118 |
| $(N,q_n)$ | Nörlund transformation | 45 |
| $\|\ \|_c$ | norm on $C_{2\pi}$ | 123 |
| $O, o, \sim$ | capital order, small order, asymptotic | 12 |
| $P(r,\phi)$ | Poisson kernel | 135 |
| $(R,q_n)$ | Nörlund type transformation | 45 |
| $R$ | real numbers | 100 |
| $R^+$ | non-negative real numbers | 163 |
| $\mathrm{Re}(z)$ | real part of $z$ | 154 |
| $S(\alpha)$ | Stolz domain of angle $\alpha$ (This notation is used in Chapter 1 only.) | 4 |
| $S(f)$ | Fourier series of $f$ (This notation is used in Chapter 5 only.) | 99 |
| $S_r(w)$ | disc of radius $r$ centred at $w$ | 4 |
| $S_n^\alpha$ | | 47 |
| $\{\sigma_n\}_0^\infty$ | matrix transformation of a sequence | 22 |
| $\sup_n$ | supremum | 23 |
| $T(r)$ | Taylor transform of order $r$ | 60 |
| $T_n$ | $\sum_{j=1}^{n} ja_j$ | 17, Ex. 7; 80 |
| $\omega_1(\delta,f)$ | integral modulus of continuity of $f$ | 115 |
| $\omega(\delta)$ | modulus of continuity | 124 |

| Symbol | | Page where symbol first appears |
|---|---|---|
| $\omega_n(x)$ | oscillation function of a sequence $x$ | 85 |
| $[x]$ | integer part of $x$ (The symbol $[\ ]$ is used frequently as parentheses. The use should be clear in the context.) | 14 |
| $\bar{z}$ | complex conjugate of $z$ | 98 |